THE PILBARA

ABOUT THE AUTHOR

BRADON ELLEM is a Professor of Employment Relations in the Business School at the University of Sydney. Since 2006, he has been Co-Editor-in-Chief of the Journal of Industrial Relations. He is a Senior Honorary Research Fellow in the University of Western Australia Business School.

Bradon's research has been published in leading international and local academic journals as well as in commissioned reports, trade journals and submissions to governments. He has written on many aspects of industrial relations, ranging from a history of the clothing industry to the impact of policy changes such as Work Choices and the Fair Work Act. For well over a decade, he was a regular visitor to the Pilbara, publishing several papers about the iron ore industry. From that long and close engagement came his book on the history of the Pilbara's industrial relations. He is now working on a major project which examines conflicts around work and communities in the natural gas industry across Southeast Asia.

THE PILBARA

FROM THE DESERTS PROFITS COME

BRADON ELLEM

UWA PUBLISHING

First published in 2017 by
UWA Publishing
Crawley, Western Australia 6009
www.uwap.uwa.edu.au

UWAP is an imprint of UWA Publishing
a division of The University of Western Australia

THE UNIVERSITY OF
WESTERN
AUSTRALIA

National Library of Australia
Cataloguing-in-Publication entry:
Ellem, B. L. (Bradon Laurence), 1956– author.
The Pilbara : from the deserts profits come / Bradon Ellem.
ISBN: 9781742589305 (paperback)
Includes bibliographical references and index.
Iron miners—Labor unions—Western Australia—Pilbara.
Wages—Miners—Western Australia—Pilbara.
Mining corporations—Western Australia—Pilbara.
Iron mines and mining—Western Australia—Pilbara.
Industrial relations—Western Australia—Pilbara.
International business enterprises—Western Australia—Pilbara.
Pilbara (W.A.)—Social conditions.

Cover image: Locomotive 5459 and train on the Mt Newman Iron Ore Railroad, December 1969. Stevenson, Kinder & Scott Corporate Photography. Ref: 340169PD.
Typeset in 11 point Bembo
Printed by Lightning Source

uwapublishing

Contents

INDIAN OCEAN

Port Hedland

Dampier
Karratha
Wickham
(Cape Lambert)

COASTAL HWY

1

WEST

NORTH

Fortescue

Millstream
Chichester
N.P.

Yule River

Fortescue River

RANGE

Key

— Highway
— Main road
...... Railway
⊗ Mine site

⊗ **Pannawonica**

HAMERSLEY

Mt Bruce ▲

Tom Price
Mt Tom Price ▲ ⊗

Karijini
N.P.

⊗
Paraburdoo

Mt Whaleback ▲ ⊗ **Newman**

0   50   100   150 km
Approx. scale

The Pilbara
Towns

NT

QLD

WESTERN

AUSTRALIA

SA

NSW

ACT

Vic.

Perth •

0   1000 km
Approx. scale

Tas.

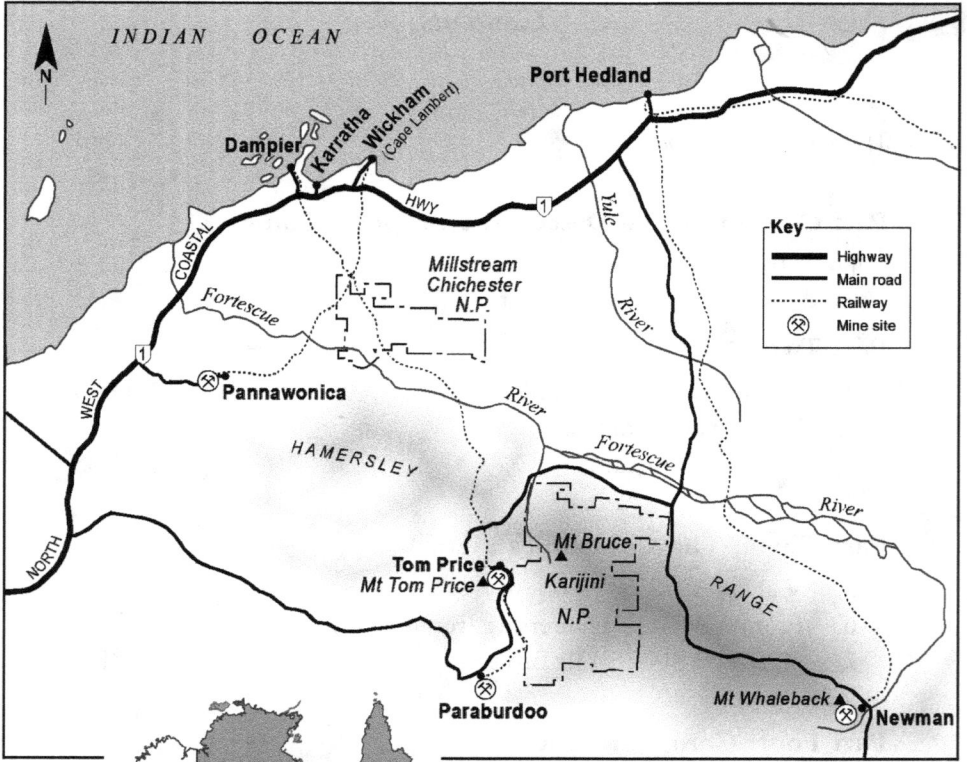

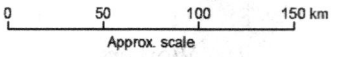

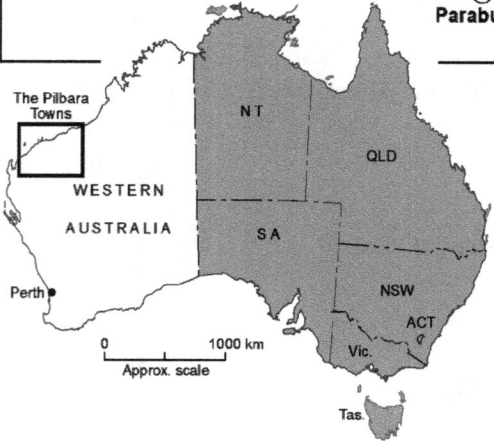

The Pilbara's mines and town in the 1980s. Map by Peter Johnson

'Welcome to the Pilbara'

Early on the morning of 26 June 2001, I got off a plane at the airport serving the mining town of Newman, setting foot in the Pilbara for the first time. My plan was to write an academic paper about how changes in employment relations were playing out in different places in Australia. With its massive iron ore sector, the Pilbara seemed an important place to examine. My paper would have just one small section on the Pilbara, then it (and I) would be back to urban areas. I never wrote that paper. Instead, the Pilbara itself became the story. Like the many people who say they are going to a mining region for a few weeks and stay for years, I was hooked. There was much more to say than a few paragraphs could hold.

At my first meeting with local workers over breakfast at a roadhouse on the fringe of the town that first morning, I was pretty nervous, carrying the dual burden of being an academic and from 'the east', knowing that plenty of so-called experts had winged in and out of the Pilbara over the years. 'Welcome to the Pilbara' was, though, the encouraging first-up response I got when introduced that morning. Over the next fifteen years, I kept returning to the Pilbara, doing more listening than talking, more learning than teaching.

Before working in a business school, I had been trained as a historian, so my instincts told me that to understand the Pilbara in the twenty-first century, I needed to go back at least as far as 1966 when the first exports of iron ore were shipped from Port Hedland to Japan. And from very early on, I knew I wanted to write this book to explain the Pilbara's history and importance.

A few weeks later, on a second trip, I had plenty of time to think about what made the Pilbara what it was. The collapse of Ansett Airlines forced me to drive the 1,600 kilometres from Port Hedland to Perth in one day to catch the only flight I could back to Sydney.

'You'll do it easy', I was assured. Well, maybe not that easy, but it was worth the long solo drive to feel the weight of the Pilbara's geography, its place in the west and something of its history. In 2004, I published a short book mainly for the workers themselves about the companies' de-unionisation plans. I promised that I would then write a fuller account. It has taken me longer than it should have, but here it is.

1. Aboriginal rock art along the Mt Newman Iron Ore Railroad, c. 1975. Norman Leslie Smithson. Ref: 140681PD.

2. Robe River crossing with iron ore reserves in the background, nd. Photographer's name unclear. Ref: BA364/702; copyright Department of State Development.

3. Port Hedland, 1972. Richard Woldendorp. Ref: BA364/489; copyright Department of State Development.

4. Houses in Mt Newman, 1973. Photographer's name unclear. Ref: BA364/560; copyright Department of State Development.

5. Living in a new house in Newman, 30 April 1969. Aerial Surveys Australia.
Ref: 267446PD.

6. Gangers laying track at Nelson Point (Port Hedland), 30 Jan. 1969.
Aerial Surveys Australia. Ref: 267374PD.

7. Drill and blast, Mt Newman, 1970s. Photographer's name unclear.
Ref: BA364/611; copyright Department of State Development.

8. Ore truck maintenance, Mt Newman, 1973. Richard Woldendorp. Ref: BA364/626; copyright Department of State Development.

9. Iron ore blasting at Paraburdoo, 1978, Richard Woldendorp. Ref: 216099PD.

10. Loading a truck with iron ore, Newman, December 1969. Stevenson, Kinder & Scott Corporate Photography. Ref 326437PD.

11. In the control room, loading iron ore into rail wagons, Newman, December 1969. Stevenson, Kinder & Scott Corporate Photography. Ref: 326458PD.

12. Engine driver in the cabin of a Mt Newman Mining Co. locomotive, Pilbara, December 1969. Stevenson, Kinder & Scott Corporate Photography. Ref: 326431PD.

13. Bucket-wheel reclaimer moving iron ore at Point Samson, c. 2001. Evan Collis. Ref: 137181PD.

14. Loading iron ore on East Intercourse Island, near Dampier, c. 1998. Richard Woldendorp. Ref: 216142PD.

15. Mt Whaleback, Western Australia, 1976. Photographer unknown. Ref: BA887/2; copyright Australian Broadcasting Corporation.

16. Port Hedland, 1976. Photographer unknown. Ref: BA887/3; copyright Australian Broadcasting Corporation.

Part One:
From the Deserts
the Prophets Come

Chapter One

Mining the Pilbara

Hoping, if still from the deserts the prophets come
(A. D. Hope, *Australia*)

THE PILBARA REGION, in the north-west of the Australian continent, has long occupied a special place in the Australian imagination. Today that imagining is more marked than ever before because the Pilbara also occupies a central place in the economy. Despite the mining boom tailing off from 2012, there are still staggering tonnages of iron ore being railed across the Pilbara and then shipped to steel mills in Asia. The industry's influence seems to be all around us: it produces up to a quarter of the country's total export income; generates fabulous revenues for shareholders in otherwise uncertain economic times; wields political influence via its companies over taxation and labour law; and gives the media a staple story on the fly-in-fly-out (FIFO) worker.

For all this, the Pilbara is not visited by most Australians and, partly for that reason, not well understood. The nature of the place is complex and surprising. Even amid the country's biggest-ever mining boom in the early twenty-first century, the Pilbara iron ore industry employed only about 60,000 people in a national workforce of 10 million. Most of the mining workforce lived outside the region; the total population of the Pilbara was only about 50,000. If most of us are uncertain about what the place is like, we also know little about how it became so important, about life and work in the Pilbara, and about how the iron ore industry affects the rest of the country. The companies, global giants such as Rio Tinto and BHP Billiton, are familiar enough in broad terms, as are the names of some of the entrepreneurs – the charismatic leader 'Twiggy' Forrest and the controversial owner Gina Rinehart, who was, at the height of the mining boom, the richest person in Australia. Apart from media talk about FIFO and salaries, what do we know about the work itself, the workers, or the place?

The Pilbara, which, for all its economic importance, employs relatively few people and seems so remote, has had remarkable effects across the Australian landscape. Understanding work and its regulation – 'employment relations' in short – in Australia's major export industry is important in itself, but there's more to it than this. Employment relations in the Pilbara have totally changed over fifty years of mining, and have deeply influenced other parts of Australian working life and politics. The Pilbara has also exemplified other social transformations: in few places can the dispossession of the Indigenous peoples and the legacy of that be as stark; in few places has the workforce been so solidly masculine and yet now so open to change, with new technologies likely to transform how mining work is done; in few places have unions risen so high and fallen so low. If the commercial history of the Pilbara has been a wild ride in the last decade, then so too, for at least a generation, has its politics.

In examining the local development and national impact of the Pilbara's iron ore industry, many different approaches could be taken and many questions asked. Most of the (very few) books written about the Pilbara have focussed on the companies, the founding figures and the leading entrepreneurs. This is not the case in this book. Rather, the work itself, the workers and employment relations are put at the centre.

To make sense of all this, the account in this book is historical, looking at the changes over time in the Pilbara up to the present day; and it is geographical, looking at what makes the Pilbara both distinctive and influential.

That this book is a historical account is obvious enough. It traces changes in the fifty years since iron ore was first exported in 1966. However, something should be said about how this kind of history is written. It is based on the idea that the past does not stay the same. This simply recognises that different people remember the same event in different ways, or that people remember events in ways that match what they think and want now, rather than what they thought at the time. So, the past changes. And these changes are not innocent either. People use and abuse history for their own purposes. In the last few years, we have seen furious debates about Australian history, most of which have been driven by people not very sympathetic to the way in which historians go about their careful work. These same problems crop up here, in trying to assess the decisive events, especially the major industrial disputes, which reshaped the Pilbara – and Australia – in the 1980s and 1990s. More broadly, the chapters in this book reflect on the common view that the once-powerful iron ore unions got what they deserved. What exactly did workers and unions do wrong and right back then? Either way, how history is read and remembered is not just 'academic'. It matters today. Companies and lobbyists routinely cast the 1970s as the 'bad old days' in their arguments for keeping unions at bay. Stories still abound of workers on strike to get a greater variety of ice-cream flavours in the canteens.

My premise that this book also takes a geographical approach needs some explanation too. One starting point is to say that places do not stay the same. What did, and what does, the Pilbara really mean to those who work there? Is the Pilbara just a mining region – simply holes in the desert connected with long-distance railways? Is it 'Rio Tinto territory' or is it a series of communities? Is it a macho place or a family place? To answer these questions, and tie them to working life, we need to think about how complicated geography is in the Pilbara.

The Pilbara is almost universally referred to as an isolated place, an extension of the isolation said to characterise the state of Western Australia (which is, in turn, an extension of the whole country's isolation). If the Pilbara is *physically* isolated, then in *economic* terms it is anything but that. It lies at the heart of global production networks, linked to the Asian steel mills buying its ores and to mining companies listed on stock exchanges in Britain and funded from all over the world. Its wealth puts it at the heart of political networks criss-crossing Australia. Thinking about the Pilbara in this geographical way begins to explain why it is so important to the rest of the country. We also get a sense that globalisation is not 'out there'. It is a local process as well as an international one. In fact, the Pilbara exists as a mining site *because of* what we now call globalisation. Without global demand, finance, labour, and shipping, it would not exist as it does, no matter how many billions of tonnes of ore lie in and below its hills.

Mining regions such as the Pilbara are at the heart of what makes the Australian economy (and society) unusual compared with other 'advanced' or 'post-industrial' societies. Australia has similarities to those countries in Western Europe, North America, and to nearby New Zealand. It has a feminised service sector employing more people than any other kind of work and, despite a cultural attachment to the 'outback', Australia has long had a highly urbanised population. However, the nation is still marked by its eighteenth- and nineteenth-century origins as a settler society: it remains massively dependent on countries elsewhere, mainly through resources – iron ore chief among them. Mining in general, and iron ore in particular, has extraordinary political significance, arising from, and also perpetuating, a geographical unevenness in our society.

What does all this mean for the Pilbara itself? If we accept that the Pilbara is a physically isolated (and often very harsh) region, but that it is bound to the global economy, then we can begin to see that this tension between local isolation and global integration creates different ideas of the Pilbara for different people. At its simplest, this is about the differences between sites for mining and places for living.

For mining companies themselves, the geography of the industry is problematic. For all their size, wealth and power, global mining companies face a basic dilemma: they cannot shift the minerals to another place if local conditions, the workers, unions or governments cause them problems. They are not as 'footloose' as other kinds of companies. So, for these multinational giants, what happens locally and nationally is vital to the success of their operations. In addition to that, mining often takes place in areas remote from major population centres, with, initially, few local workers or infrastructure.

Since the mid-1960s, the companies mining the Pilbara have had to solve these geographical problems. This book is, among other things, the story of the different solutions they have come up with. Those solutions – what geographers call 'spatial fixes' – were challenged by others, notably mining unions, and none was ever all-encompassing. What those fixes meant for work, workers and politics – not only in the Pilbara but across the country – is central to this story.

This book is organised around the three different spatial fixes which define the history of the Pilbara's iron ore industry as it transformed from a union place, to the site of a company fightback, and then to a globally oriented company space. At each step, employment relations were absolutely central to making the Pilbara what it was and is. For the first twenty years of mining, employers accepted unions and built mining towns. It is not too much to say that workers and their unions and communities 'fixed' the Pilbara as their kind of place. Over the next twenty years or so, all this unravelled as the world of work changed in and beyond the Pilbara; this was a period of intense conflict. Since then, the employers have redefined the Pilbara in their own terms through FIFO, changes to labour law, and changes to the geography of work itself. Each of the three core parts of the book explains this argument, with the Pilbara's working people at centre-stage.

This argument for why geography is significant takes us back to the question of what we do and do not know about the Pilbara. If we are concentrating on how the work of mining is done – on how the men and women of the workforce are organised to keep the ore being mined and shipped – then what happens in employment relations is at least as important as understanding the financing, leadership and marketing of the industry. Controlling work processes and dealing with the unions have been central to the ways in which the mining companies have solved their geographical and organisational problems.

In examining how the companies did all this – how they tried to control work and workers in mining – we will discover what happened to unions in the Pilbara, how a union heartland become an all-but-union-free zone, and how these developments affected other parts of Australia. What happened to work and workers in the Pilbara sums up many of the fundamental changes in working life in Australia over the last generation. It also shaped changes in the politics of work across the Australian landscape. Having said this, it may still seem odd that unions are central to this book. After all, barely 5 per cent of the Pilbara's mining workforce now belongs to a union, even less than the national average. As we will see, though, the major companies have spent a lot of energy to keep unions out, to make it seem a simple fact of life that the Pilbara is a non-union place. In truth, the battle over unionism has been a defining feature of the Pilbara iron ore story. The first thing we must do, then, is explore the times and terrain in which the Pilbara iron ore industry was something quite different from what it is today, to see how it once became a union stronghold.

Part Two:
A Union Place

In the 1960s and early 1970s, transnational mining companies, working with government backing and global finance, came to the Pilbara to service growing Asian markets. They faced a physical geography which remains central to contemporary imagination: a harsh, arid and physically isolated place; a desert outpost in an otherwise highly urbanised society. There were, at that time, no inland towns and very little labour with any mining experience. In this forbidding setting, the companies had to organise a production process to locate, extract, blend, load, rail, unload and ship iron ore in massive quantities. And this was not all. They had to build towns to house and sustain the growing workforce. These new towns quickly became vibrant centres of a distinctive social and political life. Along with the workplaces themselves, the towns became the sites of a powerful local union movement – a movement which the companies more-or-less tolerated for a generation. How did this first 'spatial fix' come about? The answer lay in part in what had gone before: for all the novelty of exporting iron ore, the Pilbara itself was no new place. Companies and workers had come to a place with long and complex histories. Those histories shaped the emergence of the Pilbara as a mining site.

Chapter Two

'Tethered to the World'

THE PILBARA COVERS over 500,000 square kilometres, across the north-west of the state of Western Australia, from the Indian Ocean to the Northern Territory border. Its physical geography has helped to make it a tourist centre; it has a stunning terrain of gorges, natural pools and plateaux inland, and seemingly endless plains nearer the coast Its physical remoteness is striking: the only city of any size in the state, the capital Perth, lies 1,600 kilometres to the south. As many have observed with a mixture of pride and dread, that city, in turn, is one of the most isolated one can imagine: the nearest city of comparable size on the continent, Adelaide, is more than 2,600 kilometres away. Perth is closer to the Indonesian capital city of Jakarta than it is to the Australian capital, Canberra.

In the 1960s, the Pilbara was almost as economically and socially isolated as it was physically remote. Mining and its transport infrastructure would change all that. As Henry Lawson had written of nineteenth-century mining, 'the mighty bush with iron rails is tethered to the world'.[1] The Pilbara would become central to global production networks in mining and steel, thoroughly integrated with international finance and logistics. In describing this transition, many see the origins of the

Pilbara's mining industry as a simple frontier story, as if nothing predated the exploration of the leases and construction of the first mine sites. Yet the Pilbara has a truly ancient and enduring history. The oldest part of that history, the geological era in which the ore bodies were laid down, is the very reason for its contemporary shape as an export mining centre of unparalleled wealth. The Pilbara's human history is at least 40,000 years old because the place has been home to Indigenous Australians for that time. The Pilbara clans, and those in the Western Desert proper, were among the last to feel the impact of the continent's colonisation. During the first period of white intrusion, from the 1860s, the Pilbara was recast as a site of sheep and cattle-raising, part of an already internationalised pastoral industry within the British Empire. Each of these three histories – geological, Indigenous and pastoral – shaped the Pilbara before the iron ore industry developed.

The Pilbara's rocks

The Pilbara owes its very being as a set of globally connected mine sites to something even older than its human habitation: the millions of years of geological development which produced some of the largest and highest-quality iron ore bodies on the planet. It also owes its current formation to the growth of steel-hungry economies in Asia, something that in historical terms is only a moment old.

Geologists were describing the Pilbara as 'iron country' with 'enough to supply the whole world' by the early 1890s, but also arguing that the ores were not economic – 'of no value' – partly because there were no coal supplies for steelmaking nearby, and partly because of seemingly insuperably long distances from any existing railway lines or deep water ports.[2] Limotite and, for steelmakers, the more valuable hematite would be revealed as the ore bodies were more closely explored and assessed. In some cases, these bands of ore ran for hundreds of kilometres, varying in thickness and quality, but at their best of rare (potential) commercial value.[3]

These formations are among the oldest on the planet, laid down from more than 2,500 million years ago from the Archean and Proterozoic eras on. The hematite in the Pilbara ore body was naturally enriched as less valuable constituents were leached out of the rocks, leaving unusually high proportions of iron content – up to 70 per cent.[4] Across these immense geological periods, the iron ore deposits were made visible (and relatively accessible) because they were less likely to be eroded than the rocks around them.

White men later named many of these more exposed deposits as mountains, such as Mounts Tom Price, Goldsworthy and Whaleback. Later still they would be better known not by those names, far less by their Indigenous ones, but by company names. Scattered across the Pilbara, these mining blocks ranged from older and smaller ones near the coast in the north at Shay Gap, to huge deposits in the south, some a mere 20 million years old. Tom Price and Whaleback, now mined by Rio Tinto and BHP Billiton respectively, were two of the most obvious and richest examples. They presented as prime sites for open-cut mining. Still immensely profitable, these ore bodies, which had been millions of years in the making, will likely be mined out in just one century of capitalist intervention.[5] Viewed in this light, the often-asked question of what we gain from mining booms assumes a wholly new dimension. Posed as a question not simply about material gains, profits and taxes, it becomes a question of ecological significance: what will this rapid stripping of millions of years of natural development mean for us in generations to come?

Indigenous history in the Pilbara

Between the aeons of geological development of this potent but transitory resource and today lay thousands of years of human habitation. The place which became the British colony and then the state named Western Australia was inhabited and marked by human beings for at least 40,000 years, with perhaps 12,000 people living in and around the Western Desert on the eve of colonisation, in as many as thirty-one different language groups.[6]

Towards the end of those years, in the seventeenth century, the Indigenous Australians living in and near what we now call the Pilbara became aware of white men and their ships. These first whites to see the west coast came in search not of minerals but of spices. In the seventeenth century, these men came and went. The Indonesian islands were of much more interest to them than anything in or near the Pilbara.

The first steps in the process which transformed the Pilbara were the steps taken onto the shores of Sydney harbour, 3,500 kilometres away, in January 1788. In establishing the site for a penal colony, from which, within a generation, a sprawling pastoral economy serving an international wool trade would emerge, the British began to lock the Australian continent into what we would now call the global economy. In doing this, they were caught in a dilemma: these potential pasture lands were inhabited by people unwilling to relinquish them. The contradictory solution was to declare the land *terra nullius* ('nobody's land'), but also to offer Indigenous people the protections afforded to all British subjects of the Crown. The former allowed massive dispossession and, in fact, near genocide; the latter could see, at least in New South Wales, white men hanged for the murder of black men and women. However, beyond the reach of the law, beyond the growing colonial towns, a little-recorded war took place, especially where the pastoral industry expanded. The squatters, as the men taking up the huge leases on offer were known, had an all-but-insatiable demand for land and, more importantly, exclusive use of it. This could mean nothing other than conflict with the Indigenous people, for whom land was not merely material but also spiritual.[7]

The British did not establish a colony in the west until 1826, and for many years that colony was mostly confined to the south-western corner of the continent. Only from the 1860s did the whites begin their push into the north of the colony. From the beginning, this invasion, too, was all about pastoralism. Over the next sixty to seventy years, it was as brutal a process as it had been in the east. One set of reasons for this was that as the nineteenth century wore on the whites' ideas were shaped more by 'social Darwinism' than by any acceptance of Indigenous life. Whites were superior; there was no such thing as the 'noble savage'; the benighted first inhabitants of Australia were a dying race. 'Terra nullius' now trumped rights for black-skinned men and women. The other reasons concerned the economic geography of the west. Pastoralism was less encumbered in the west than it was in the eastern colonies by state intervention, or by small farmers seeking to challenge the squatters. Paradoxically, unlike in the east, the pastoralists were forced to rely upon Indigenous labour because the colony's rulers forbade the use of convict labour in the north.[8]

Until the colonisers changed their lives forever, the people of the Pilbara had enjoyed lives under conditions at least as prosperous and healthy as those of white workers in Europe. The Pilbara's economy was a sustainable one, based on sharing food and materials; only a short time was spent in hunting and gathering to secure daily foods. No one family or group, far less a class, controlled any surplus that might be produced. Whites struggled to comprehend the social structures and practices and spirituality by which these people lived, and which sustained their daily existence.[9] Many physical expressions of that ancient culture remain, most famously in the form of the petroglyphs on the Burrup peninsular, which are among the oldest remaining examples to be found anywhere.

The rise of the pastoralists

From the 1860s until well into the twentieth century, the Pilbara was the site of effective rule by the squatter; the white landholders (often in the form of British companies) more or less did as they pleased. For them, the only question was whether the 'natives' should be impressed into labour, attacked or left to die out. Despite its formidable resistance, Indigenous society did not survive the struggle with the pastoralists and the colonial state intact. Nonetheless, Indigenous inhabitants such as the Martu were among the last to abandon nomadic life, and came into missions or cattle stations as late as the 1960s.[10]

Killings on the Burrup Peninsular in the first half of 1868, recalled now as the Flying Foam massacre, set the tone for frontier violence with the arrival of the pastoral industry. Thereafter, the loss of lands, and then the forced removal of children, dislocated Indigenous life well before the mining booms of the twentieth century.[11] The pastoral economy also drew, however, on local skills, much more so than mining, in employing (if that is the correct word) many Indigenous workers. The colony's legislation, in particular the inaccurately named *Aboriginal Protection Act, 1886*, lay behind the conditions of workers on the pastoral stations. The pastoralists had to provide food, some minimal clothes and health care, but the law said nothing about wages. Noel Olive argues that this was, in effect, slavery.[12] There were few countervailing forces because the police, often brutally, aided in the physical control and movement of the local population, while even the church missions were kept out of the Pilbara. Other forms of labour were no less demanding, not least the dangerous and brutal life of pearling into which many were forced and which, in at least one account, sustained the pastoralists' incomes through tough times on the land.[13]

Despite this mixture of repression and hyper-exploitation, there were ways in which the Indigenous peoples managed to survive and adapt. The most important of these – often to the alarm of the whites – was that people walked back to the fringes, if not the heart, of the Western Desert. This form of sanctuary lasted longer here than most other places on the Australian continent. The other adaption over time was that the skills picked up in pastoral work could be turned to other purposes, including non-wage labour, and that many families made their own use of the small mining sites which had opened before the iron ore industry developed. They worked as prospectors and small-scale independent miners themselves.

The pastoral industry became the scene of one of the most extraordinary industrial disputes in Australia's history when Indigenous workers walked off the job on May Day 1946. Two Indigenous men, Clancy McKenna and Dooley Bin Bin, along with a white so-called trouble-maker, Don McLeod, organised a strike. McKenna and McLeod were gaoled several times over three years, as were workers themselves. The government denounced McLeod as a communist stirrer who had upset the otherwise contented 'natives'; the media paid the strike almost no attention, according to one account because of 'rich and powerful friends' in the Pilbara, that is, the pastoralists.[14] There was not much support for these men and their families, though seafaring unions did back them in actions as well as with words. 'Arcing up' over working conditions, therefore, predated the Pilbara's iron ore industry and the white workforce.

The strike leaders and the stockmen not only stayed true to their cause as workers and strikers, but went on to make their own distinctive history. Most never went back to work for the pastoralists. The dispute was never formally resolved. Many of the workers and their families simply walked away from the global capitalism which had been defining their working lives. They established cooperative forms of production, set up successful properties and schools, and engaged in extensive mineral prospecting. In the early 1950s, they purchased properties with the profits of some of these ventures.[15]

None of this small-scale mining and hardly any of these families became part of the iron ore export industry. It is an intriguing thought, though, that the early development of mining in the Pilbara was marked by work outside of capitalist arrangements, arising directly from a labour dispute in another industry, and was carried out by the descendants of the oldest inhabitants of the place.

The making of the iron ore industry

If the history of the Pilbara is older and more complex than it is commonly imagined to be, then the story of the origins of iron ore mining also has its surprises. The key players were located far from the Pilbara. The impetus came, as Osmar White put it many years ago, 'not from visionaries' entranced by the Pilbara as a place, but from 'industrial realists who knew nothing and cared less about the country and the people who lived in it'.[16] And then the role of Australian governments became vital.

The state government granted the first iron ore mining leases for operations on just two islands, Cockatoo and Koolan in Yampi Sound, a little to the north of the Pilbara, off the Kimberley coast. The leases were for the mills of Australian Iron and Steel (AIS) on the east coast of Australia at Port Kembla and Newcastle. Mine construction began in 1935, the same year in which AIS became a wholly owned subsidiary of one of the companies later to reshape the Pilbara, the Broken Hill Propriety Company (BHP). In Geoffrey Blainey's pioneering study of Australian mining, he describes this government intervention taking place after a Japanese company had shown an interest in the deposits – an early indication of the global networks that redefined the Pilbara thirty years later.[17]

When construction work began, union membership for workers was the norm, as was wage-fixing through arbitration tribunals. Notwithstanding the impact of the Great Depression, union membership in 1935 stood at 44 per cent of the workforce across Australia, and was slightly higher in the west.[18] That wages and conditions would be set through a union and by 'awards' in the arbitration system was more or less taken as given. The Australian Workers' Union (AWU) was the union entitled to cover these employees. Its representatives met with the Yampi Mining Company and the government's Industrial Registrar aboard a vessel called *Yampi Lass*. The negotiators settled the first Iron Ore Award in 1937. The award built on precedents set in the state's goldfields and in iron ore mining in South Australia. How telling, and how (once) characteristically Australian, that the business of industrial relations was encoded in local geography: the waters where the deal was struck were renamed Arbitration Cove.[19]

This first iron ore venture came to nothing. Not until after World War II would these small sites with their high-quality ores be mined. When that happened, there were further discussions between a range of unions (now others in addition to the AWU) and the company with detailed rules about job classifications and pay rates. When mining started on the islands in 1951, wages and conditions were already set out, as were the demarcations between unions and, with that, some aspects of how work itself would be organised.[20]

There had been no exports of iron ore from Yampi Sound in the 1930s, in part because of a dispute on the waterfront at the very site where the main BHP steelworks were located, at Port Kembla in Wollongong. Waterside workers refused to load a vessel bound for Japan, arguing against supplying the Japanese war machine which was so brutally taking control of China. Ironically, those two countries, Japan and China, became the major markets for the Pilbara's ores after World War II.[21]

Governments accepted the conventional wisdom that the country's iron ore stocks were of limited scope. Indeed, although privately alarmed by the strategic threat of steel firms in Japan, governments chose in public to emphasise the paucity of ore stocks and the need to preserve them for local steel making. Even when some people did let themselves dream of the Pilbara's potential, the geographical problems of the place's isolation and the inland location of the ore bodies appeared insurmountable.[22] One of the few to call for a rethink, with what might be called typical West Australian optimism, was a Minister for Mines who told the state's parliament in 1938 that changes to transport 'might make very inaccessible ore bodies very accessible'.[23] True enough, but for the time being the state government was as obdurate as was the federal government: it ruled that prospectors who 'pegged' out potential sites were not guaranteed any return. It offered such certainty only to companies which mined the sites and promised to develop manufacturing – a policy quietly abandoned as mining boomed in later years.[24]

For its part, the mining and steel conglomerate BHP, at that stage a purely Australian company, was also reluctant to see the Pilbara's ores mined. It leapt in to secure the leases at Koolan Island which the Japanese had wanted, largely to stop anyone else getting a foothold in the potential industry. Koolan Island and the long-used mines in the Middleback Ranges near Adelaide met the demands of the company's steelworks in South Australia and New South Wales. As late as the 1960s, the first moves to develop the Hamersley Iron company under the forerunner of Rio Tinto were opposed by BHP. No doubt by that stage the company's leadership was also looking at what use it might make of the Pilbara ahead of its London-based rival.[25] For thirty years from the early 1930s, though, two of the key players which were later to profit from remaking the Pilbara as an export site – BHP and the state itself – were determined to keep it locked up for local use only.

With no private-sector interest, the public sector stepped in. The Western Australian government took over the small existing mining leases here, as it did elsewhere in the state. The first major site of iron ore mining production was neither in the Pilbara nor off the Kimberley coast, but in the south of the state at Koolyanobbing, 400 kilometres east of Perth, towards the Kalgoorlie goldfields. From the 1950s, these mines provided iron ore to a government-owned smelter outside Perth and, thereafter, to a new AIS plant at Kwinana, south of the capital city.[26]

Unlocking the ore bodies of the Pilbara for overseas suppliers was, then, a story of politics and government-company networks, not, as the foundation or pioneer myths would have it, a story of heroic, individual men taking on nature and the world. That Pilbara iron ore industry myth is Lang Hancock's self-promoting story of his ore discovery while off-course in his small plane in bad weather over the Turner River – a discovery he kept quiet while lobbying for the export embargo to be lifted. This romantic story has been debunked in other accounts, beginning with Neil Phillipson's over forty years ago and more recently by Malcolm Knox. Phillipson's assessment of weather records, local geography and flight routes makes the story, says Knox with some restraint, 'highly dubious'.[27] In 1957, five years after Hancock's apparent adventure, another even-lower-profile discovery was made to the north, when a prospector, Stan Hilditch, who was looking for manganese, came across the prodigious ore body that would become the Whaleback mine at Mt Newman.[28]

The more prosaic story about the making of the Pilbara, as we have started to see, is about the governments and transnational corporations – which Hancock was fond of vilifying. He cleverly positioned himself to take advantage of the sites he had pegged after he won over previously sceptical Conzinc Rio Tinto Australia (CRA) executives. These were the businesses which went on to provide untold wealth to him and his family after he made extraordinary royalty deals with CRA, the company that would become Rio Tinto.[29] Hancock secured a 2.5 per cent royalty rate for as long as CRA/Rio mined the sites, and went on to make an unimaginable fortune without actually working in the industry. Complex legal manoeuvres kept these rates quarantined from change and kept the windfall in the family.[30] By contrast, others would later be resistant to paying anything over 0.5 per cent to Indigenous owners, if they paid at all.

To be clear: for anything to change, for the Pilbara to be 'opened up', state intervention was an essential condition, most obviously through the federal government lifting the ban on iron ore exports and the state government overturning its long-held policy about pegging, and to recognise those claims, as happened in 1961.[31] Once these policy changes had been made, the Western Australian government entered into agreements with mining companies about the conditions for the development of the Pilbara.[32] With more certainty about their future, the companies undertook further prospecting, and CRA in particular uncovered truly astounding prospects.[33]

The prospector Stan Hilditch and his partner, a mining engineer, Charles Warman, became involved in negotiations similar to Hancock's over their stake in the Mt Newman site. There were very protracted discussions involving state and federal governments (who themselves did not see eye to eye on the best way to invest), American companies and BHP, before the establishment of the Mt Newman Mining Company.[34]

Blainey suggests that there may also have been a short-term political expedient for the federal government in ending the embargo, because economic growth and international funds were needed to counter a recession in the early 1960s.[35] This was a time when, as happened in 1961 (strange as it now seems), an unemployment rate of 2.3 per cent nearly cost a government an election. In what became a familiar story, change in the Pilbara and, with it, the rest of the country, was driven by a mix of backroom pressure, public lobbying and immediate political needs.

With government policy changed, and with ongoing support from the bureaucracy in Western Australia, international investors and global corporations moved in. Given that, as a site of export mining, the Pilbara was new ground, and that start-up costs were massive, it is hardly surprising that state support and global finance shaped the Pilbara mining industry from the beginning. The Australian state opened the way for global giants to profit from the Pilbara's resources; but, with that done, it walked away from the attempts to use resource exploitation to drive downstream manufacturing development, as even conservative policy makers had wanted to do before World War II.

It is vital, therefore, in understanding how the Pilbara was remade as a mining site to appreciate how important governments and all their agents were in working with mining transnationals to develop the industry. A later premier said that the state was committed 'at all stages to assist developers in the marketing of their product, in their search for capital, and in their search for suitable partners'.[36] As Herb Thompson's path-breaking work on the Pilbara made clear many years ago, this was not empty rhetoric. Governments provided the mining companies with long-term leases, tax concessions and specific measures to cut through normal planning processes. The conservative state government was more than willing to fight its supposed political allies in the federal government whenever the latter threatened to complicate the development of the Pilbara. The state government's expenditure in the Pilbara doubled in the five years after 1958.[37] As construction work began, the *West Australian* newspaper reported that $1,631 million was to be spent on infrastructure, $730 million of which was to be public funds.[38]

Little wonder that critics and researchers came to use terms like 'institutional capture' to argue that, because of the narrowness of the state's economy and low levels of population set against the wealth and power of the global mining lobby, state and society in the west were hostage to the mining transnationals.[39]

The Pilbara's mining past was shaped not by one or two heroic individuals or by companies alone, but by Australia's colonial history, Indigenous dispossession, the rise and fall of pastoralism, and by governments and politics. In the end, it was not just an Australian story but an Asian one, as countries relatively nearby – and brought closer by the size and reliability of new ore-carrying ships – sought high-quality ores for their rapidly growing steel industries. From the beginning, the Pilbara iron ore industry was 'tethered to the world'. It was into this web of history and geography, of politics and power, that workers came and tried to make sense of the Pilbara in their own ways.

Work and employment relations – laying the foundations

When construction work on the mines and ports began in the early 1960s, the Pilbara's main communities were still composed of family groups in the Indigenous population, with minimal links to the capitalist economy beyond pastoralism. The other major groups were still the pastoral families and British owners who controlled the vast stations. This is not to say the old and new white economies did not overlap. Lang Hancock was the most notable crossover from pastoralism to mining, from one form of wealth generation to another. For him, mining's claims to land were at least as powerful as those of the pastoralists or anyone else's: 'Nothing should be sacred from mining, whether it's your ground, my ground, the blackfellow's'.[40] More generally, however, as had been the case in industrialising Europe many years before (and still is the case in parts of Australia), the relationship between farmers and mining companies could be fraught. As mining began, the destruction wrought in rebuilding towns and developing railways often distressed Pilbara farmers and townspeople alike.[41]

The logistical challenges facing the new mining conglomerates and their backers were formidable, as many have recorded. There were no towns near the mines. The only harbour, at Port Hedland, was too small to allow ore carriers to manoeuvre, let alone load. Inland transport links were sparse. There was no railway between the ore deposits and the ports, hundreds of kilometres away. There were hardly any roads worthy of the name. This mining development phase drove the first population boom in the Pilbara, with the number of residents more than doubling in the 1960s; the male population of Port Hedland tripled. In the same decade, the number of people employed in building and construction rose from 210 to 4,603.[42]

Because the mining companies had to lock in customers and delivery dates, there was intense time pressure.[43] The first project was the Goldsworthy operation, owned by US and Australian interests. It was the nearest to the coast, only about 110 kilometres from the shipping facilities at Port Hedland. Mine, rail, port and housing facilities were completed by international and Australian contractors within just fifteen months. The first stage of the more complex CRA Hamersley Iron project was also completed with great speed. Its open-cut mine at Tom Price, 300 kilometres of rail and its port facilities in Dampier were all built in only twenty months. Ore from that project was shipped a little after

Goldsworthy's first exports in June 1966. About eighteen months later, the company's second major mine, at Paraburdoo, was producing ore. Some downstream processing was also beginning with the establishment of a pellet plant at Dampier. In the meantime, Mt Newman Mining Company – in which BHP had a 30 per cent stake through a subsidiary – had developed its pit along with the Pilbara's longest rail link, and expanded shipping facilities at Port Hedland. Finally, nearer the coast, Cliffs Robe River, under US, Japanese and Australian control, began development work in 1970, exporting ore from 1972. A rail line connected the mining site at the new town of Pannawonica with the port near Wickham about 280 kilometres away.[44]

The workers who built this infrastructure came from all over the country, and well beyond it. Some had worked in other parts of the north; others had come with experience from the biggest publicly funded project in Australia, the Snowy Mountains Hydro scheme, which had drawn workers from many European countries. As well as Anglo workers from the eastern states, New Zealand and Britain, there were large numbers of Yugoslavs and Italians and some Thursday Islanders – even some Japanese.[45] In some respects, then, the Pilbara was forerunner to the changing ethnicity of other parts of Australia. What did not take place was any attempt to draw on Indigenous labour. Indigenous workers remained on the margins or confined to the pastoral sector.

Most of the workers were, by definition, itinerants. Their mobility and the type of work they performed made them typical of the membership of the biggest union in the country, the AWU. Through a federal 'award' (as opposed to the state jurisdiction of most other unions) and through its wide-ranging coverage clauses (as existed in its power base in another pastoral and resource state, Queensland), the AWU had the potential to organise and represent almost all the non-trades workers in mining construction. As early as 1965, the union had a full-time official in the Pilbara. His successor, Gil Barr, lived in Port Hedland from 1968 to 1972 before becoming the AWU's Western Australian secretary, the first in a long line of male union officials who would emerge from the iron ore industry to take leadership roles in Perth.[46] That the Pilbara produced labour leaders alongside its iron ore became a common theme of Western Australian industrial folklore.

Union and state-based regulation of work was the accepted norm, but it did not sit well with some people more used to a looser way of doing things in the Pilbara. One American construction manager recalled that the 'New Australians' did not mind hard work and laboured away 'without the thought of a strike entering their minds'.[47] While, then and since, managers have often assured themselves that all their workers are content with their lot, the Pilbara was being transformed whether managers liked it or not. This was due to the changes being wrought by global employers as they opened up the Pilbara for mining – industrialising a previously pastoral site and necessarily introducing new types of work and workers.

It was, in part, the geographical peculiarities of the Pilbara which began to drive this change and, in particular, allowed the newly unionised workforce to win improved wages and conditions. For trades workers in the Pilbara as elsewhere, an award for the metal trades was the foundation for wages and conditions. Precedents had been set for construction work on the North West Cape in 1964. The 'Cape Award' mandated messes and accommodation as well as wages higher than those in similar awards and in so doing played, on one account, 'a significant role in the introduction of over-award wages and conditions in the iron ore industry'.[48] The benefits came at some cost to workers. Overtime was built into the 'normal' working week, which was, in effect, fifty-four hours. The rationale for, and implications of, these new norms were complex. The immediate interests of employers seeking to meet contractual deadlines coincided with those of workers keen to make money as quickly as they could in the harshness of the Pilbara.[49] Soon enough, the interests of employers and workers diverged, and the wage militancy of workers would cause concern not only to mining employers but also to some union officials. In the mid-1960s, however, there was little thought of such things.

Intriguingly, the desire to have some order and regulation in the emerging resource sector disrupted standard practice. Ray Fells explains how the state Employers' Federation had undertaken all the bargaining for employers (even the usually independent BHP), which had led to 'a strongly disciplined local scene' built around awards for three key (male) industries, metals, building and transport.[50] This uniformity, arguably unique to Western Australia, did not survive employment growth in the Pilbara; the shift to the agreed longer working week at the Cape was the start of this change. The Pilbara iron ore industry was disruptive from the beginning.

The state Industrial Relations Commission agreed that if workers wanted the money for working in such harsh places and if employers wanted timelines to be met, then that was fair enough. This view, coupled with the growing presence of the AWU and other unions in a place with high labour demand, set the scene for wage competition between employers. In this rapidly changing setting, iron ore construction and mining was launched.[51]

Although unionism was the norm in blue-collar industries, and arbitration and awards were broadly accepted as the methods by which standards were set, this does not mean that work and conditions were not tough or that getting workers unionised was straightforward. Many obstacles confronted the unions: the very newness of the sites and the workers, high labour turnover, ethnic divisions, the isolation of the worksites within the Pilbara and of the Pilbara itself and, at times, rivalry and outright hostility between unions. The pressure on the small number of union organisers (and their families) based in the Pilbara was intense because workers, themselves often isolated and lonely, demanded prompt action on their grievances.[52]

Just as has been the case more recently in construction of gas industry facilities in the Pilbara, accommodation was a source of much conflict. Cramped and hastily partitioned quarters made for one set of complaints when four men shared a space of about nine square metres. Then, as now, shift changes disrupted rest and privacy. Food was generally considered above average quality, but when Poon's catering was being commonly referred to as 'pooman's power packs', the workforce was obviously unhappy with how the heat turned the food from being just fine into warm and soggy sandwiches.[53] In few places in Australia was industrial action taken to have the Esky replace the plastic bag for storing lunches. The Pilbara was one.

The emergence of iron ore mining in the Pilbara

From the beginning, nothing was as it seemed in the Pilbara. This supposedly new and fresh place was in truth built on geological formations which were aeons old. This apparently isolated site and archetypal Australian space assumed its modern form because of global interconnections. The Pilbara's much vaunted and even mythologised white frontier and individualist culture was built on the importance of the state – from its roles in ignoring the violence towards, and dispossession of, the first peoples in the nineteenth century, to its detailed intervention in underwriting the activities of transnational corporations and investors in the twentieth.

The new mining sites being created out of the Pilbara's vast and ancient spaces were as different from older mining sites as could be imagined. The coalmines of Europe and the Australian east, and for that matter in Collie in the state's south-west, were typically close to ports or industrial areas, and often had pits in close proximity to each other. In folklore, if not always in practice, these kinds of towns were sites of a political militancy which mirrored the politics of the mining worksites and provided the strategic base for more generalised radicalism across a national working class.[54] In contrast, the mines of each company in the Pilbara were hundreds of kilometres apart. With that, so were the towns – isolated not just from Perth but from each other. The distances covered by any one company could be just as striking, most obviously in Mt Newman's case where the mine and town were 400 kilometres from the end of the rail line in Port Hedland.

This geography provided problems for the mining companies in developing and sustaining production, and it also constituted a spatial threat to workers' solidarity that was quite distinctive compared with other mining regions. The well-established transnationals and unions both brought their own locally specific answers to these problems as mining began to flourish, with the unions at first making the Pilbara seemingly their own, as the next chapter will show.

Chapter Three

Frontiers of Control

IN THE FIRST six years or so of mining, from 1966 on, the industry boomed and the Pilbara quickly became home to a particularly active set of unions, with highly developed formal networks at the worksites and informal ones in the towns. Strike levels were among the highest in Australia; male wages came to be well above average.

None of this 'fixing' of the Pilbara's geography and life was uncontested. Many of the workers and local union leaders hoped, and other people feared, that a new kind of unionism might be built in the Pilbara, delivering to workers not only high wages but control over the sites in which they worked and the towns in which they and their families lived. The companies, too, sought to impose their own ways of ordering and making sense of work in this place. Nonetheless, in the Pilbara, far from the formal seats of corporate, union and political power in the metropolitan centres, the workforce did create a distinctive unionism and, with that, a distinct sense of place.

How this militant and apparently powerful form of local unionism emerged in this physically isolated part of the global mining industry has been the subject of myth-making as much as careful examination. In trying to rectify this, we must recover this period from what, in another context, the English historian Edward Thompson referred to as 'the enormous condescension of posterity'.[1] In other words, putting the working men and women who made their own history at centre-stage – looking at the Pilbara as they saw it and tried to change it.

Mining the Pilbara: Work organisation

To some people, mining looks like a simple business, merely digging up and moving dirt. To others, the only thing which might make it interesting is the current focus on automation. For both the sceptics and 'techies', labour, understood as workers or the work process itself, is not very important. Yet the question of how work and workers are organised is vital because, as in all sites of production, the labour process turns potential work into actual work, and value is created. In the Pilbara in the 1960s and early 1970s, work organisation was controversial, lying at the root of high levels of conflict and at the heart of what made the place distinctive.

Managers tried to organise work so as to minimise unit costs and integrate all parts of the production process, from drilling and blasting to loading and shipping the ore. Just as the construction process had been shaped by tight timelines, so with mining itself: the challenge was to get 'the dirt' to the ports as quickly as possible to meet buyers' demands. This was about two things: control and cost. Capital-intensive from the beginning, the companies wanted maximum utilisation of that capital through economies of scale, be it the huge 'benches' in the pit (fifteen metres high at Whaleback), the big shovels, the ever-bigger trucks or the two-kilometre long trains. These defining features of the Pilbara iron ore industry were often (and still are) talked about in purely technical terms, and often in awe-struck tones (so to speak). And they are undeniably physically impressive. The underlying reason for this scale was, however, neither technology for its own sake nor engineers amusing themselves with big toys. It was a means to one overriding goal: better use not only of that capital but of the associated labour.[2]

The mining labour process in these massive open cuts, some of which would soon be three to four kilometres long, was structured around long-term planning based on initial drilling and mapping to calculate the best methods to mine each block in the pit. A manager's challenge was, as one summed it up, to coordinate and control 'from a bore in the ground...to...the marketplace'; every aspect of production, from exploration for specific grades of ore to ship turnaround times, was critical.[3] The industry quickly moved, as the next chapter shows, from trial and error to computer-based technologies for planning purposes.[4] One fundamentally important point is that what labour-process scholars call job 'conception' (planning) was carved out from 'execution' (doing); the iron ore mining process was, in its own way, as fragmented a set of jobs as mass manufacturing had become.

No-one has better explained the tensions which flowed from these kinds of work arrangements than Carter Goodrich in writing about conflicts in Britain during the post-war upheavals of 1919. His book's title, *The Frontier of Control*, became a standard term to describe what he called 'workshop politics'.[5] This is precisely what was at stake in the Pilbara's workplaces. These struggles started as soon as mining began, mirroring wider conflicts around the world as, in the wake of other political and social struggles in the 1960s, the idea of 'workers' control' came into its own.

The Pilbara employers organised their operations in distinctive ways. As we shall see, their first solution to the problem of control was simply to pay high wages and more or less hope that this would resolve workplace tensions. In other mining sites, for example Western Mining's Kambalda nickel mines in the south of Western Australia or the Broken Hill metalliferous mines in New South Wales, workforce consultation was given more importance; conflict was channelled through accepted union structures. Those companies sought (or had imposed upon them) close and often fruitful relations with the local communities in the mining towns.[6]

The labour process was broken into very distinct sections, with different types of work and skills at each stage. Drilling and blasting were the first steps and, in the late 1960s, characterised, according to some, by the 'error' in 'trial and error' because of the haste with which work was done, and an unsophisticated approach to getting high-grade ore. The blasting damaged men's ears in the early years, and at times parts of the mine sites too, because setting up explosives was often imprecise, as was the radio communication.[7] Once blasted, the ore was loaded by shovels into trucks. The ore was tipped into the primary crusher where two workers were on duty, one as operator, one as cleaner. The work was, like all the work on the mine, dusty and potentially hazardous because large rocks could block the crusher. Blasting also took place in the primary crusher, requiring the men to climb in and out and, as with the blasting on the hill, to get the charges right. By accident or design, this did not always happen. Some sites had facilities for pelletising the ore, a process designed to add value by increasing the burn temperature and purity of the ore. The pellet plant added a significant degree of complexity due to the maintenance needed: electricians, skilled fitters and boilermakers were essential. From the outset, much of the work process was mechanised and in massive proportions. Loading the trains at the mine and then unloading, 'reclaiming' and ship loading at the ports was an era away from the old shovels and small trucks of underground coal-mining. The rail operations needed not only drivers and yard workers but also workers with high levels of maintenance skills as well as large numbers in track gangs.[8]

At several links in this production chain, key workers could exercise power over the labour process. They could disrupt the whole production system: shovel operators taking the ore off the face, workers in the power stations, fitters or electricians doing emergency maintenance, train drivers shifting all that tonnage, ship loaders overseeing the ports. This was one of the weaknesses managers faced in the system they had created. On the other hand, the detailed divisions of labour made it difficult for the workforce as a whole to act together and to challenge management's general power over them.[9] Stockpiling ore served not only commercial purposes but, the companies hoped, stood as a bulwark against industrial action by workers.[10]

These categories of work were codified in the first mining award for the Pilbara, made in 1967. It set out six distinct functions: quarrying, crushing, transporting, treating, storing and loading.[11] The tribunals thereby gave legal sanction to the division of labour which the companies had created. The award process purported to regulate workplace relationships collectively, that is between unions and their employers. At the same time, it codified the fragmentation of the workforce. In years to come, unions often threw themselves into bitter disputes with each other about who had the right to cover which types of work.

As in the construction phase, no-one doubted that the award system would set the minimum wages and conditions for this new industry. This meant that unions would be accepted as parties to the regulatory framework. Because of union coverage rules, the unions were for the most part the same ones as in similar industries elsewhere in Australia. At first, there were twelve such unions. Three were very small: one craft union, the Australasian Society of Engineers, and two white-collar unions. There were three building unions which were relatively unimportant once development work was complete. That left five major mining unions: the AWU, with nearly half the total Pilbara membership; the Amalgamated Engineering Union (which merged with other 'craft' unions in 1973 to form the Amalgamated Metal Workers' Union, the AMWU), covering trades workers such as fitters and boilermakers; the strategically vital Electrical Trades Union; the Transport Workers' Union which, in a geographically specific oddity, had (and still has) coverage of truck drivers 'on the hill' at Mt Newman; and the Federated Engine Drivers' and Firemen's

Union (known as the FEDFA in other states), which represented shovel and crane operators, plant attendants and rail workers – all particularly critical links in the production chain.[12]

There were ideological and strategic differences between the unions, with the AMWU, or the 'Metalies', generally regarded as the most left-wing. That said, most of the unions were more militant – at least in obvious terms such as the propensity to strike – than their branches in other places. Because of this, one of the key local officials early on, the AWU's Gil Barr, was looked on warily by his state leadership. Regarded as a moderate union elsewhere, the AWU was at times different in the Pilbara. Barr later said that he 'was always a militant when…young. Everyone was. I just went along with them'.[13]

In years to come, unions spent a lot of time in conflict with each other in demarcation disputes over whose members should have access to which tasks. There were even disputes within the one union (the AMWU) about the line between the work of a boilermaker and a fitter. Routinely criticised by employers and the media for this, the unions were, in fact, dealing with underlying divisions which were not of their making, but which reflected the work structures which the companies had set up.[14]

Unionists' insistence that some jobs only be done by certain workers drove the managers crazy but it had its own logic. More than one former union leader was explicit about this: no one should do someone else's job; it was about 'protecting and creating more work', recalled John Beales, an AMWU convenor.[15] Demarcations also allowed the unions to control overtime because, of course, the strict divisions meant more call-outs for more specialist workers outside their shift times.[16] This way of thinking had a long history because of the uncertainties of working-class life in general, and was common in mining with its boom–bust cycle. Still, many unionists worried that employers could use all this to play one union off against another. Repeating the old adage that 'in union is strength' was one thing, but putting it into practice was quite another.

Setting the precedents – wages and conditions in the 1960s

The Pilbara was subject to the normal forms of work regulation through compulsory conciliation and arbitration, but these were in some ways made afresh in the 1960s. Unlike sites of new production in other places, the Pilbara had few communities or local customs around industrial work. To paraphrase Edward Thompson once again, the workforce 'was present at its own making'.[17] The workers were as new to the place as the industry to which they came, but they arrived with their own ideas. Many came from union backgrounds and understood the merits of collective action in the workplace. Others, at least in folklore, were of a more individualistic bent, on the run from failed relationships, the police or creditors. For these men, the politics of the more militant union members might have had no appeal, but the collective fight for improved conditions was nonetheless attractive.[18]

That a federal system would regulate work in the vital iron ore industry was unthinkable in the west. By far the majority of employees worked under awards and agreements in Western Australia's state system throughout this period.[19] At first, when mining began at the Goldsworthy and Hamersley sites, work ran ahead of regulation: there was no specific award to regulate the industry. The AWU's general award and the metal trades state award provided the framework for the first year or so of mining.[20]

The establishment of an award system to regulate wages and conditions in the industry was driven in the first instance not by the unions, as might have been expected, but by the employers, organised under the state's Employers' Federation. In June 1966, Western Mining Corporation (which was mining iron ore at Koolanooka, near Geraldton) joined with Goldsworthy and Hamersley to seek an award for the industry, based on awards for Yampi Sound and the South Australian industry. Hamersley was the driving force in seeking this regulatory intervention because its managers wanted a more stable system to try to reduce their very high levels of labour turnover, then running at 180 per cent.[21]

The unions asked Commissioner Kelly, who oversaw the early hearings in the state's industrial commission, to reject the rationale of the employers' claim and to use instead the higher paid quarry workers in the eastern states as his benchmark. They could not persuade him to do so. Nor could they set a precedent by winning their claim for an industry allowance to allow a share in 'emerging prosperity' and to compensate workers for the problems associated 'with the industry and the areas in which it functions'.[22] The phrasing of this claim was interesting. It anticipated arguments which academic geographers would later make that work and place, industry and geography, cannot be untangled from each other.[23] After all, all jobs are performed in a physical place with its own peculiarities.

The award made few changes from the conditions under which metal mining operated elsewhere. Yet the Pilbara *was* different: in other places, awards and agreements governing work were part of a social fabric of established communities, whereas the Pilbara sites were isolated and, more to the point, barely existed as towns at all. Living conditions were different – and this plagued attempts to regulate Pilbara iron ore as if it were just another industry.

As was standard in Western Australia's unionised sectors, there was a clause providing for 'preference for unionists'. On taking a job, non-unionists had to apply for membership of the union with the right to cover their work and had to remain a union member while employed. This was, in effect, compulsory unionism. Unionists' long-held view that workers reaping the benefits of collectively won conditions should not be 'free-riders' was thus satisfied. This was not, however, the legal logic of the provision. Rather, it was designed to allow the objects of the arbitration acts to be met, namely, to encourage collective representation as a means to achieving industrial stability.[24] This practice remained substantially intact until attacks by the employers at Robe River in 1986 and Hamersley Iron in 1992. Of similar importance, although nothing came of it, was the unions' claim to have a clause in the award which would have formalised the role of shop stewards.[25] This claim arose from the Pilbara's geography: in such a setting, argued the unions, a 'proper employer/employee relationship' required union organisation 'on the job'.[26] Rather ironically, this thinking anticipated the employer push from the 1980s for 'enterprise bargaining'.

This first award had a term of just one year. It was intended to be an interim arrangement as the industry developed. And develop it did, with the workforce nearly doubling in 1967 to 1,134. That number more than tripled in the next three years.[27] The new mines were quickly becoming large workplaces. This fact, coupled with the interconnections within the labour process, meant that management could not simply set things up and then let them run. Re-planning and making fresh decisions was the standard, which opened the door to the view among workers that processes and outcomes were unfair – and that, ultimately, managers were always in the wrong.[28]

Hearings for a new award did not commence until late in 1968. This time, Mt Newman was to be a party to the award. All the Pilbara operators and their workforces would now be covered by it. Wages and leave formed the unions' main demands, together with the industry allowance, this being of the 'utmost importance' to the unions.[29] A local AWU official recalled that the workforce generally did not like the first awards, and when his members and others went on strike, the state leadership 'thought they had all gone off half cocked'.[30] The first awards therefore not only failed to bring harmony between workers and employers but also saw the first break between locals and union officialdom.

The major changes which flowed from these hearings were that the unions were allowed to make claims for an industry allowance, and that grievance procedures were set up which allowed the shop steward a role.[31] Easy as it is to read this as an example of the growth in union power, it is a much more intricate story than that. It was the employers who wanted some such procedure (preferably with more teeth than the version in the award) to secure, in Lovett's words, 'some level of formal control over the growing power of the site union structures'.[32] Could the companies regain control by sharing it?[33] As Herb Thompson succinctly put it, the very workplace structures the companies had insisted on actually enhanced the power of shop stewards against them and the state union leaderships.[34] The union officials in Perth had their doubts about the grievance procedures and shop stewards, seeing them not as a challenge to managerial power but to their own.[35] It was envisaged that the companies (not the unions) would train the stewards, arguably because union headquarters were 1,600 kilometres away from the mines and ports. With that geographical disruption came a different mental world: the courts and meeting rooms of Perth were far away, in every sense, from the heat and dust of the Pilbara.

Increasingly, local unionists, not the state officials, began to drive improvements in working conditions. Shop stewards and the 'convenors' they elected as leaders from their own ranks combined and pushed for fresh demands while using the grievance procedures to assert their power over daily matters of workplace routine. Ray Fells drew on the term 'reluctant militants' to describe at least some of these workers.[36] They may have had union experience from another site, then when an issue arose they would be put forward to take up the case with managers. If they did well, they were stuck in the role.

Soon, there were specific agreements at Goldsworthy (for a bonus scheme) and Hamersley (following industrial action over dust). The employers were caught in a dilemma as workers leveraged improvements at one site to make demands at another.[37] The award was not as successful as its architects had hoped. The closest study of this process argues that connecting local wages to national standards only added to instability in the Pilbara when there were changes in relative pay rates in those other awards.[38] Because these awards were minimum rates awards and were linked to wage changes in other sectors, localised bargaining became more, not less, important – facilitated, of course, by employers' demand for labour, the normalisation of overtime and workers' power at critical parts of the labour process.

By 1971, the employers were looking for a way out, a return to a system cf more uniform standards across the Pilbara; but, like the state union leaders, they knew that the site unions had to be involved.[39] Just six years after the first iron ore exports and after a series of failed attempts to regulate work in the Pilbara, the employers, the Commission and union leaders in Perth were looking for a new way to control work and workers. This begs the question of just what was happening on the ground.

Work and militancy

That the Pilbara became the site of a militant unionism seems to be taken as inevitable, but there was no reason why this would necessarily have been the case. All the apparent causes – isolation, high labour turnover, the free-spirited nature of some workers and the short-term monetary focus of others – could, in principle, have led to apathy or outright opposition to collective action in general – and strikes in particular. The rise of the Pilbara unions, then, was not predestined. It needs to be explained.

The failure of the regulatory system to bring 'order' to the Pilbara provoked dismay among the players in the negotiation process. In folklore and even detailed research, attempts to understand the place were often solely based on geography (harshness and isolation), the assumption that the behaviour of the workforce was deviant ('immature' was a common description), or simply the newness of the industry to the west. For some in the companies, it was the 'ego needs' of the convenors that created that environment; for others in the media, the pace of development and the demand for labour meant that 'the Pilbara got off on the wrong foot'.[40]

The Pilbara can be read in other ways. When a former Pilbara resident, asked to say when she first went there, answers 'let me see…it was just before the '88 [Newman] strike', a researcher knows that the rhythms of personal life are recalled in very particular ways. These local and class-specific features are starting points for understanding life as these men and women experienced it. For many people in the Pilbara, this place was becoming the new norm; it was other places that were different.

Workers' own accounts of the nature of their work and of life typically stress that although mining wages rose to be well above average, living and working conditions remained harsh. These conditions and, on some sites, heavy-handed supervision of the workforce, were among the most important sources of individual discontent, shared senses of grievance, and more-or-less spontaneous collective action.

The climate, physical terrain and labour process made for a tough job in every way. Moving between worksites or simply trying to get out of the towns for a break meant driving on mostly new but invariably poor roads, some more like tracks. Dust and heat were overwhelming on the mines in particular, but also at the ports. A westerly wind blew red dust all over the buildings at Port Hedland where the loading facilities were adjacent to the old township. Long-standing Hedland residents were horrified at first.[41] The housing for the growing workforce would be built at a new settlement of South Hedland, about twenty kilometres away from the dirty port.

Workers arriving in the Pilbara were stunned by the heat and the flies. As one, Graeme Stevens, put it, recalling his 1970 arrival in Paraburdoo: 'Standing outside at 4 o'clock in the morning and you were picked up by flies. It was terrible...the drive in was terrible...The road was creek beds all the way in'. The following year, John Purdue arrived: 'I didn't know what I'd struck when I first got here...It was all cobblestones and rough and dirt and dust and Landrovers and 4WD's going past...I thought "God, what have I landed in here?"'.[42] Roger Parsons, later an AWU leader, recalled that when he arrived from the south of England, via Perth, he was shocked: 'All of a sudden the road finished, at least the bitumen did...I didn't know whether I'd lost my way or what the story was. I'd never seen anything like it before'.[43]

Despite the high turnover in general, these men stayed, as did so many who had planned just a few months in the Pilbara. They explained this in terms of social life, improved amenities and what today would be called family-friendly rosters and working arrangements around eight-hour shifts. Bob Dickson, a foreman, recalled: 'there was a terrific team spirit, or community spirit'.[44] Dickson thought the best thing to happen was the hiring of married men because men alone, be they married or single on their own, got into alcohol habits that were hard to break. John Mossenton (later state secretary of the AMWU), however, was troubled by the strain on marriages and the 'eye-opening' number of affairs in the town of Dampier.[45] The ones who stayed and who were active in the unions provided the core attitudes which shaped the Pilbara amid the generally high labour turnover. They were starting to set the workplace norms.

On the job itself, the pressure to meet customer demand in this harsh environment put the unions in a good position to increase wages, but the very same circumstances drove the demand for long working hours and often chaotic work organisation. Workers on the Hamersley mines recalled that low-grade ore 'was just dumped in a stock pile...they hadn't been stripping waste and overburden' so the high grade ore most readily available was being quickly depleted.[46] Long days were worked. The work was usually organised around three eight-hour shifts – so it was continuous production from the beginning – but a sixty-hour week was immediately established as the norm. At first, then, it was hours and overtime which drove earnings up, not high hourly rates of pay as such.[47]

Machinery was a major source of discontent: 'in the early days, those drills made me deaf', recalled one worker. At times, the noise from the Tom Price mine reached as far as the town. On the mine, there was little muffling. 'Bloody horrendous', was one summary. In a truck's cabin, 'supposedly sealed', not only was there no air conditioning for the hot days, there were sometimes no windows. On cold nights, 'you'd have rags around your head trying to keep yourself warm'. As for 'the dozers, they were just open. Blokes used to stand between the blade and the radiator to try and keep warm'.[48]

Workers were at first sceptical that anything could be done about the physical conditions. As one later said: 'when the unions started pushing for airconditioners…and things like that I thought they were having a lend of themselves…Anyway, they did, and now, of course, they wouldn't be without them'.[49]

Strikes led to much consternation in the Pilbara and a bemused tone to the scholarship seeking to understand them. It is quite clear, though, that the basis lay in the tough conditions, often intemperate management and also a good-humoured and larrikin culture which was emerging. Inconsistent supervision also featured: on the very same site, workers doing the same jobs would be treated differently from one another. Workers later said that one foreman would work with the men about overtime, another would say: 'I've had you bastards…you will not get overtime'.[50] One foreman at Tom Price 'caused more strikes in Hamersley than any half a dozen foremen put together…he just came to work spoiling for a fight and nine times out ten he got accommodated'.[51]

All the available statistical evidence points to dissatisfaction with management as a major cause of disputes.[52] Relationships with the companies were complex. On one hand, the bosses 'had all the power… [they] could sack a bloke…really at whim'; but, on other the hand, most interactions were with the foremen, who were 'pretty close to the men in those days'. Workers might be able to get an under-performer sacked – at least one foreman confirmed that slackers were not tolerated – but there was a lot of confusion: 'I think', said Dickson, 'that was the start of the militancy of the union movement. The blokes wanted…some consistency'.[53] Doug Alchin, a fitter at Tom Price, summed up management strategy nicely: 'everything was tons driven…and they tried to buy industrial peace'.[54]

The very fact that so many workers were driven by short-term monetary concerns – and that at Goldsworthy the mine was expected to have a short life – only intensified problems. The nearby town of Shay Gap was widely considered one of the better places to live, but even here tensions arose. John Bryant, a manager who worked there and at several other Pilbara sites, provided an acute analysis. Explaining why a worker would smile one day at being called a bastard by a foreman, but the next day want to walk off the job, he said:

> It was…a pressure cooker…both in the living and working situation…there not being much housing and they were working long hours…They would literally work themselves into the ground, they were short of temper, the facilities… were less than would normally be required for that number of people. So the shop was crowded, the pub was crowded, so it was a pretty uptight organisation.[55]

Two things were clear: high levels of conflict were all but inevitable; the workers believed that, through the unions, they were the only ones who could push companies and tribunals to change conditions.

Conditions may have been problematical, but there is no denying that the first years produced, at least in the retelling, a can-do and good-humoured culture. The companies would later clamp down on all this in an effort to improve performance by effectively bureaucratising it. These kinds of stories are too numerous to recount, but among the typical accounts are the story of the backhoe that was demolished at Tom Price when a blast went wrong with the result that the operator got a new and better one. He was delighted with the result and also no doubt at not being in the backhoe at the time. When another blasting miscalculation sent a rock from the crusher through a workshop roof, this was, understandably, enough for the electricians inside: 'They just turned out the door – in the bus and went home'.[56] As for the morning-long water fight with fire-extinguishers at Tom Price at Christmas 1974, that would, said one of the participants, mean 'marching orders' if it happened today.[57]

In these circumstances, what did workers and their unions actually do to reshape the Pilbara? There was an overlap between living and working conditions, union growth and militancy. For although the awards required workers to belong to unions, it was still the case, certainly at first, that many existing workers were not union members. As is still often the case, the main reason for this was that no-one had asked them to join. Parsons was explicit about this: after noting that, early on, disciplinary matters were often sorted out by 'a punch up',[58] he went on to say that joining a union was simply a matter of course. The only reason he had not joined up before coming to the Pilbara 'was because I couldn't find how to join'.[59] That union membership was required was not a problem. Local action, not top-down bureaucracy, made them into union members of a particular type. So much is clear from the account provided later by the AWU's Gil Barr. Between 1968 and 1971 the union's membership increased from 700 to 4,600. Like other local officials later, he found the work pressure relentless, driving hundreds of kilometres a week, often getting up at 3am to drive to another site for a meeting at 7.30am. Interestingly, in his view, collectivism preceded formal unionism: 'the blokes...always used to stick together but weren't in the unions'.[60] Addressing physical conditions, not least the dust, was no less decisive in winning workers to the union cause.[61]

For Barr, the key issues were managerial control over the workers' very livelihood. Workers often joined the union in disputes over what they saw as the arbitrary dismissal of one of their mates. The early awards included a suspension clause which the workers hated but, said Barr, the managers 'used to love'.[62] Workers could be suspended for two to three weeks for making a mistake. Only when workers started suspending themselves in support of others did the companies ease off. Many other workers recounted how they saw it as being vital that they challenged managers at this 'frontier of control': 'we took that [managerial] right away from them...You couldn't survive any other way'.[63] A dismissal could see a worker set down on the highway with 'a bottle of water and a few sandwiches' to wait for a lift back to Perth.[64]

The union convenors and shop stewards were, for many, the essence of all that was wrong in the Pilbara, the cause of all the problems. Interestingly enough, in the very early days, at least one of the Perth-based union officials had the same reputation. The high-profile AMWU organiser (and later state secretary of the union) Jack Marks was accused by Commissioner Kelly of stirring up trouble. He said he always knew where Marks was when he went north 'because of the trail of trouble he left behind him'. It was as if he was 'dragging a burning bag' across the Pilbara.[65] By the same token, however, many workers in the south of the state had taken to calling Marks 'Back-to-Work-Jack', as he sought to mediate disputes and control the more militant workers. So, then as now, the role of union leaders at all levels was complex and was about managing different perceptions within unions as well as with employers.[66] The union convenors were indeed powerful figures, but they themselves believed that the source of that power was the union members. Formally, this was obviously so: workers elected their shop stewards who in turn appointed the convenors.[67] More than this, convenors could not afford to be out of step with those workers. Managers were always troubled that the pay systems seemed to encourage workers to 'blow' Friday and make up earnings with overtime the next day. Stories of men turning up to strike meetings with their fishing gear already in the car were common.[68]

The speed with which workers decided to go on strike alarmed people just as much as the volume of strikes. Barr's account of one dispute with the Mt Newman company, in this case over bonuses, sheds some light on this. After overtime bans and short strikes had achieved nothing – 'the company just laughed' – the workers had been on strike for three weeks. When Barr reported to members in Port Hedland that there was no change in the company's position, the workers promptly decided to stay out another week. In Newman, someone asked if it was true that Port Hedland had dealt with the issue in nine minutes; the 400 people who had turned up voted, this time in four minutes, to stay out. Commissioner Kelly was unimpressed, but Barr pointed out there was actually nothing more to report and that mass meetings were not the only way by which members and leaders communicated. So, a four-minute meeting was easily explained, albeit with tongues slightly in cheek.[69]

Men and women, mines and towns
The construction period had laid the basis not only for the production of iron ore, but for the production of practices and cultures around work and community. A core element of this storyline was a masculinised Pilbara 'wild west' story. Ross Calnan, later an ETU convenor at Goldsworthy, told an interviewer that in the 1970s, the Pilbara really was:

> like a wild west camp…you worked fairly long hours…No single women working in the mine so you can imagine all you really did was work and have few beers after work, then you've got time to do your washing and have a kip, and that's about it, then you're back at work.[70]

One of the first major strikes, at Hamersley's pellet plant, was marked by an apparent threat to blow the plant up. The foreman took it seriously and spent the night on guard, working his way through a case of beer and shooting the empty cans with a local policeman's .38 revolver. No bombing took place; the supervisor, on his account of the night, was kindly allowed to go home and sleep if off when he turned up at 6.30 am.[71] The fact that the workforce was almost wholly male reflected wider social practices which had constructed mining as men's work and which had built a particular kind of masculinity around shared, hard – and union – work. So, there was a macho ethos on the worksites which seemed to play out in militant unionism, with aggressive personal confrontations part of the site cultures. Perhaps unsurprisingly, there was a high turnover of HR managers in the Pilbara, more so than among other managers.[72] The only 'real men' among the managers were ex-workers for whom there might be a grudging respect. For the most part, a particular masculinity was a weapon against managers, just as it was a means to exclude women from the union culture.[73]

None of this is to say that the Pilbara was ever solely a male domain. The companies quickly came to favour a workforce of married men, reasoning that this would provide more social and industrial stability. In the Goldsworthy towns, workers believed that single women had been hired 'to hold the single guys in the town'. To Calnan's amusement, the pub fights then stopped and 'in the club...there was aftershave and suits on'.[74]

Segregation remained marked, though: single men's quarters were built as utilitarian sites (they still are), while houses for families were built around them. There had been Indigenous women in paid labour in most sectors; there were white women in the small services sector. On the mining sites, just as the blue-collar jobs were cast as men's work, so the companies reinforced Australia's marked gendered divisions at work in hiring women to care and clean – and type. As families moved in and towns developed, a 'non-mine' labour market was created which was mostly female in teaching, health and town administration.[75] Much of women's labour was invisible, notably prostitution, shrouded in the curtains of gender and racial denial. And the working lives of women in the Pilbara were often as dull as working life in the lesser skilled mine jobs.[76]

As the industry and the towns began to develop, companies not only passed over the workers who would have been the most immediate source of labour, but excluded them. The companies agreed to measures that would not disrupt the pastoral industry's access to Indigenous labour, so the iron ore workforce was almost wholly white. On one account, Indigenous employees and residents in or near Newman were moved out.[77] The towns would be as white as the workplaces.

The new towns across the Pilbara were literally company towns. Under the agreements made with the mining companies, the state government required the companies to fund and build the infrastructure for the towns on the land leased to them. Tax concessions and reduced mining royalties more than sweetened the pill for the companies. Town and industry were thoroughly integrated: these were closed towns, de facto rule ceded by the government to the mining companies. This meant that all the public utilities were also controlled by the company. The companies used low rentals on their air-conditioned homes as part of their strategy to attract labour at a time when wages were still not that high. Subsidised power and water completed the home package. The development of shopping centres, sports fields and public parks slowly transformed the desert into a kind of suburbia.[78]

Closed company towns they may have been in law, but not necessarily in fact. By the late 1970s, the companies were only too happy to have them 'normalised' with local government control. One reason for this was the costs; another was that they had become sources and sites of conflict. As at least one senior manager at Hamersley realised, the solidarity and discontents forged in the labour process overflowed into town life: 'issues don't die on the job', he said.[79] There has been much debate about the impact of company towns and of how, in particular, town issues flowed into workplace discontent.[80] What is most significant, though, is this: by far the most important (if not the only) collective voice in each town was the local union movement. As the workforce grew and the population of each town expanded, so did the range of issues which the unions took up. Accommodation, along with schooling and healthcare, were major sources of concern for workers and their families.[81] The unions raised local concerns ranging from the need for dentists, to televisions, to better pubs for families at the Robe River sites.[82] These concerns came up in all the towns, but it was widely remarked that there were differences: the smaller sites and towns of the Goldsworthy operations were generally considered better and friendlier places to live. Newman, by contrast, was often seen as a tougher town with more obvious social divisions between workers and 'staffies'.[83]

The overlap of town and mine issues was clear in Port Hedland, where the local Labor Party worked with the unions to campaign over grocery costs. Carol Fagan, a school teacher and Labor Party secretary, organised pamphleting of the town, and uncovered how prices were being driven up. An active consumer group was set up, prices monitored and, of course, fresh wage demands made.[84] While many deplored the extent of union power in the towns, the scholar then closest to the unions argued quite the reverse: that the unions had paid too little attention to issues like housing and, later, normalisation, as against 'getting a few more dollars in the pay packets'.[85]

Whose Pilbara?

The basis for union power and, with that, the remaking of the Pilbara lay in a combination of processes that were specific to time and place. Harsh and isolated geography, tough working conditions and employer strategy set the scene in a new industry. Questions about work and conditions that had long been contested elsewhere were all up for grabs in the new iron ore industry.

All this can be understood as a necessary set of circumstances in explaining the nature of work, employment relations and unionism in the Pilbara in the late 1960s and early 1970s, but that is only a start. What did these workers and their unions actually do in these circumstances? That unionism would flourish was not a certainty; the form it would take still less clear. An emerging, shared set of values and practices among this otherwise disparate and highly transient group of workers drove these people to make something of the place. The particularly harsh working conditions and the pressures for continuous production gave a special edge to the Pilbara. Then, as the towns developed into community spaces, they reinforced and were shaped by the workplace power of the unions. By the early 1970s, the workforce had gained a much greater control over the production process than had others across the country. For the time being, 'outsiders' – be they tribunals, state union officials or employer representatives – struggled to assert much say over the Pilbara.

For all their success, the unions had weaknesses. A sense of community, through sporting competitions and combined union organisations, might have reduced fragmentation across the Pilbara, but did not overcome it. The worksites remained integrated more by the employers' production system than the workers' own organisation. One study went so far as to say that there was really no 'strong workplace union organisation'; merely, 'autonomous sectional industrial action'.[86] A pro-union researcher's take was perhaps the most acute: the gains workers were making obscured the limitations to them. Unions were strong, but members needed to appreciate 'the cyclical nature or foundation of that power'.[87] From the mid-1970s, the global and national contexts would become tougher, threatening the conditions the unions had won, and the way that the Pilbara was understood.

Chapter Four

Contested Terrain

FOR MOST OF the 1970s and into the early 1980s, work and employment relations in the Pilbara were marked by a less-certain global context than that which existed between 1966 and 1972. Affluent societies such as Australia's felt the effects of the oil crisis, and more general threats to the high levels of profit and employment which they had enjoyed since the end of World War II. In the Pilbara, these changes did not play out until the mid-1980s; but, on the eve of that period, as ore sales to Japan slowed and market competition from Brazil increased, Robe River and Hamersley Iron introduced workplace changes that were signs of things to come.

It is sometimes said that the state of product markets determines most things in employment relations.[1] In particular, during downturns workers are more cautious and might abandon collective action altogether as employers encircle them with cost-cutting and pressures to work harder. The Pilbara workers, however, read the changes they faced from market, employer and government pressure as more complex than they seemed at first blush. True, the demand for labour eased off and managers attempted to introduce workplace changes, potentially undercutting the unions' overall position, but it was also the case that profits and productivity were still increasing. Even when governments and police stepped in, the Pilbara unions were reluctant to back off.[2] Workplace discontent remained unabated and, despite the growing challenges they faced, the workforce, their ever-growing communities, and their unions were still able to define at least some of this 'contested terrain'.[3]

Control at work

After the sometimes haphazard way in which work had seemed to be organised in the first years of mining, the organisation of work quickly became better planned. To achieve a rapid and cost-efficient flow, managers entrenched the division of labour and their control of the technical side of production. The geologists' and engineers' work was now aided by computer-based technologies. Mathematical modelling drove the extraction of ore from particular parts of the site to meet production and shipping schedules at the biggest pit, Mt Newman, from the mid-70s. At Hamersley, planning was similarly centralised and computerised: blasting plans, the nature of the ore and the precise position of each shovel in the pit were closely integrated. All of this was underpinned by planning, years ahead, the process of extracting ores of specific qualities to meet the needs of different sales contracts.[4]

In recent years, there has been a lot of discussion about automation, some of it giving the impression this is all new; but, as Kosmas Tsokhas' 1986 account of work in the mining industry shows, there was an 'overwhelming trend' towards computerisation which involved 'the appropriation of...know-how' by the 1970s.[5] In his summary, there was a shift in power from worker to manager which 'enriched mental rather than manual labour'.[6] From the design of pits to the details of ore extraction, 3D graphics were playing their part from 1981, and some tasks, notably preparing explosives, needed minimal involvement. Workers were monitored through two-way radios when mining on the hill. Managers at Hamersley made clear how things were changing, reporting in 1980 that trucks were 'directed to specific shovels by a radio despatcher to optimise truck utilisation. Truck availability of 80 per cent is achieved by rigid pre-start checks, driver training programmes and extensive planned maintenance programmes'.[7]

These control process were a double-edged sword for the companies: they reduced production costs and workers' power in the labour process, but still did not guarantee managerial power. Things did not always go smoothly, to say the least – supervisors and workers alike complained about rusted and unreliable equipment, especially at Finucane Island and Goldsworthy's smaller and, many felt, under-capitalised operation.[8] Tensions around control of production still generated conflict. As the managers at Mt Newman came to realise, they had created a labour process which, for many workers, lacked inherent satisfaction and was 'almost mindless'. Specific groups of workers might experience what seemed to them to be random demands to speed up or slow down, to work more or less overtime, because of problems somewhere else in this complex and integrated work system.[9] The unions still exercised control over aspects of the rosters and working-time arrangements. The companies, faced with unions and arbitration tribunals that were locked into the employment relations system, had not (yet) contemplated control over one other aspect of their operations, the direct control of pay rates.

For all the famed (or notorious) power of the Pilbara workforce, the employers' control over technology was difficult to challenge. The unions' militancy was chiefly around wages and conditions and work-time arrangements. Attempts to regain influence regarding technological change were fraught, often ending up in disputes between unions. Some workers did push for a fundamental challenge to managerial power, but most of them settled for extracting the maximum possible material gains from their employers.[10]

After the 'award period' to 1972, the strike wave in the Pilbara only became more pronounced. On any measure, there was an upward trend for most of the 1970s.[11] Most strikes were workplace specific, involved a small number of workers and were (at least superficially) quickly resolved.[12] Strikes were not the only form of industrial action. Because capital utilisation and throughputs were so critical, the workforce knew that overtime bans and other limitations were, as Tsokhas put it, 'useful tactics in disputes with management'.[13] Officially recorded causes of strikes varied over time and between sites, but by and large the major stated reason between 1967 and 1980 was managerial prerogative (in 495 of 1,174 recorded strikes). In terms of time lost, however, wage disputes were much more important as they involved more workers or ran longer (or both).[14]

That the 'typical' strike involved fewer than fifty workers and lasted under two days reinforces the perception that the frontier of control was being constantly tested. This was mostly so in the two biggest operations, Mt Newman Mining and Hamersley Iron.[15] Faced with this high level of strikes and its locally specific features, experts and company officials alike were as bemused as they had been in the late 1960s. According to one careful study, managers had, by the mid-1970s, 'become accustomed to strikes and come to regard many of them as inevitable'.[16]

The companies did not change their approach to managing work. Rather, they fell in with the common fallacy that their problems were all about troublemakers – a view which the state government shared, or at least appeared to share. This meant that the solution seemed to be to attack the convenors and shop stewards, the assumption being that the workforce would suddenly see them as a problem and not the means to solving their problems.

Few features of the Pilbara's union scene attracted more media outrage than the rise of convenors. These 'war lords' seemed to sum up all that was wrong about the Pilbara, with a supposed iron grip over the worksites. Realities were more complex. Convenors were not paid officials like the union organisers based in Perth. They were elected, usually by the shop stewards, and were not paid a union salary. In some cases, they were allowed by the company to do only their union work and so became, in effect, a full-time union officials but ones whose wages were paid by the company. There were inherent tensions in having companies pay for a union role. These reflected wider contradictions in the nature of the role – and its possibilities. Some convenors simply came and went from the post but some became union officials and others 'went over' to the company, becoming supervisors. If a convenor saw the role in purely instrumental ways, as a step towards a higher position, then either path might be taken. If the path to a full-time union job 'down south' was blocked, there was always another option – becoming a supervisor and taking the route to a management job.[17]

While in the role and if truly committed to it, convenors came under competing pressures. They could be fierce advocates for their members, but this could land them in hot water with their union leaders as well as employers.[18] On the other hand, some convenors were thought to be out of touch with workplace concerns and 'soft' on the company. One group at Goldsworthy did away with the position because of that remoteness.[19] In short, there were pressures on the convenors to conform as much as to disrupt. Backed by the members, the position, and the people in it, could be powerful. Over time, as later years would reveal, it could, however, be no threat at all to employers, but merely another layer of bureaucracy.

From 1972, the role of convenor began changing. The first full-time paid convenor was probably the AMWU's Frank Wagner, in Hamersley's locomotive workshops in Dampier. For the same union, Norm Marlborough (subsequently a Labor politician) assumed an almost legendary status in this role from 1975 at Robe River's Cape Lambert site. When the company began to have second thoughts about the role and started to withdraw some of his privileges, his members promptly went on strike. Others were no less troublesome to the company, not least the AWU convenor, Dave Dawkins, who outraged the Commission as much as he delighted his membership with his focus on local needs as he saw them.[20]

No account makes clearer the nature of the work done by the convenors than Colleen Heath and Chilla Bulbeck's intimate and moving story of life in Newman in the 1970s, in *Shadow of the Hill*.[21] Heath's husband, Chris, an Engine Drivers' convenor, was closely involved not simply in the much discussed stop-work meetings and strikes at the Newman site, but also in something usually overlooked, the provision of emotional and practical support to union members and their families. It was draining and almost ceaseless work: men and women would telephone the Heath home at any time, not least before and after the eight-hour shifts, regardless of Chris Heath's own schedule. More formal business was never far away either: an evening out would often come to an end with a phone call saying that one work group or another had a problem and could Chris let the next shift know.[22] With pressures mounting and illnesses in his family, it all became too much: Chris Heath took his own life. The aftermath made all too plain that there was a dark side to solidarity in the community: his widow was blamed for his death and assaulted.[23]

The most revealing and, for local workers, positive aspects of the way that Heath and Bulbeck tell the story is that the Pilbara was not so much a 'different' place as a new norm for those living there. Localised attitudes to work, the company and unions were coming to be taken for granted. This was especially so in relation to attempts to bring union organisers into the Pilbara. Even this idea could not overcome, for Colleen Heath, the fact that they were still not part of the community: 'You cannot bring a person from the outside to go to the men and say what they should do and what they shouldn't do'.[24] As with the much-criticised localism, so with militancy in the Engine Drivers' Union: rather than being led by the nose by unscrupulous convenors, most of the members were keen for information, keen to be involved precisely 'because it was a very militant union'.[25] There were intense debates not just about wages and conditions but politics and power in the industry. Policies were not imposed; members were closely involved in developing claims for new agreements. Running disputes in the Pilbara did have its fun side, as many said, allowing more fishing time for some, but it was not just a lark. With three shifts a day, with different panels and crews, the more dedicated convenors and shop stewards were flat out running between different parts of the site to hold meetings and gather and convey information.[26]

Disputes were not confined to the workplace. The towns themselves were still sources of discontent. Most of the discussion of normalisation focussed on the apparent decline of company control, but at the same time that the companies were reducing their (tax deductible) costs, they were also hoping to take the sting out of the social issues in the towns. Many people – researchers, managers and union officials alike – often acted as if community issues were unimportant. Logically, this was untrue – workers needed somewhere to live and they needed to be rested and fed to be able to work. In the Pilbara's iron ore industry, these social considerations were obviously vital because of the single-employer status of these isolated towns. With an almost all male workforce, something very traditional, the nuclear family, was vital to keeping the mines going. At this very time, gender structures were becoming unsettled across the country in places with more complex and diverse local economies as the number of women in the paid workforce grew and campaigns for equal pay and better conditions followed. Less so here.

The towns were already changing before normalisation. The gender balance was one indicator of this. There were still very few women working in the industry, but from the mid-1970s, the rate of increase in the female population ran well ahead of men's. Norm Dufty reported that 'the masculinity ratio fell from 151 per cent to 137 per cent' between 1976 and 1981.[27] Families were typically young and large; schools and sporting clubs began to flourish. Yet, the towns remained company funded and company run, with most of the married workforce living in rented houses, others in 'quarters'.

Unions and, paradoxically, communities themselves had little say in the process which was driven by government and companies.[28] Local reactions, especially around the shift to home ownership, were complicated. In the forty years since then, it is probably true to say that despite the falling affordability of housing, the idea (or ideal) of home ownership has become even more entrenched in the psyche as desirable. The shift from rents to mortgages and the precise nature of the arrangements under normalisation generated a lot of debate in the towns. So too did the wider question of what it might mean for unionism itself. 'Buy a Home and Sell your Soul' was the headline for an article on this in the AMWU's national journal.[29] A generation earlier, a Queensland politician had argued that home ownership would 'promote greater industrial harmony', precisely the thing worrying those who argued against it because it would mean 'a more docile, submissive labour force'.[30] There was actually very little discussion of the issue in the local unions. True, they were given no formal opportunities to intervene, but this had not stopped them being active in other fields. Their focus remained on the politics of the workplace, but the companies and the government were shifting the social landscape around them.[31]

If union control within any one town or worksite was not always as robust as it seemed, it was still less marked across the Pilbara taken as a whole. Despite much discussion about setting up a Pilbara-wide political organisation or labour council, neither ever emerged. One veteran Pilbara union official, Denis Day, who favoured a labour council, argued later that the failure to set up such an organisation had 'robbed the workers of a general staff to coordinate the struggle...against the employers of labour'.[32] Unlike Day, most of the people pushing the proposal were from outside the Pilbara. Their proposals seemed designed to exercise power over the local workforce, not power for them.[33] In other mining sites (most notably Broken Hill), powerful regional labour councils had emerged from the bottom up. This never materialised in the Pilbara.

Regulating the Pilbara

As high levels of industrial conflict continued in the early 1970s, the leadership of the four companies and the state officials of the mining unions came to a shared view that a new way was needed to regulate wages and conditions and, at the same time, to bring the workforce to heel. There were five rounds of company-specific awards from 1972 to 1979, aiming to calm things down. Despite challenges from both sides, this set of arrangements lasted, at least on paper, until the 1990s when, one by one, the companies dismantled any form of union-based bargaining. Others have provided detailed accounts of these awards, so here the focus is on what they tell us about the nature of local work and employment relations, and how wages and conditions changed overall in the 1970s.[34]

In the negotiations in 1971 the employers once more tried to organise themselves collectively, this time under an industry-specific employer association, the Australian Mines and Metals Association (AMMA). AMMA's move into iron ore negotiations after the Employer's Federation had taken the early lead merits a brief comment. The organisation was founded in 1918.[35] Low-key at first, AMMA became more forceful a generation later as the leading proponent of non-union, individual agreements.

Unity between the companies was short-lived in the face of very difficult negotiations, with the unions refusing the companies' final offer; the underlying reality was that different work practices had grown up at the different sites.[36] There were work stoppages in December, after which Hamersley split off from the other three companies to make a separate agreement which conceded higher wages than had been in the supposed 'final' offer. For some classifications, the increases were over

20 per cent on the rates under the 1969 award. The agreement was registered in March 1972. Hamersley saw the process in a positive light, writing it up rather optimistically as a development of 'mutual trust'.[37] With the precedent set, the other companies followed with their own agreements. Wage differences between the sites were reduced.

One thing the agreements did not achieve was a reduction in strike levels. The practice of paying high wages to try to secure industrial peace did not succeed. Obviously, there were problems with processes as well as outcomes. Once the companies realised this, an additional agreement was reached covering grievance procedures, stand-downs during strikes, and the formal recognition of combined unions committees (CUCs). The CUCs would be made up of the shop stewards and convenors on each site. They were to meet with company managers fortnightly. This agreement ultimately failed, largely because conflict continued unabated.[38]

When negotiations began again in January 1974, the employers had re-united, seeking standard wages and uniform over-award rates. They also wanted to remove production bonus schemes in Mt Newman and Goldsworthy, replacing them with a scheme based on seniority – presumably to encourage labour retention – at all four companies.[39] Differences in rail haulage distances, production systems and rosters compelled the creation of four different agreements, but the negotiations were designed to produce agreements with as many terms in common as possible. Fifty-eight site representatives attended the meetings in Perth, a worrying feature for the full-time union officials not only because of the sheer number but also because of their propensity to intervene and explain, often forcefully, to those in the room just how things worked (or did not work) in particular workplaces.[40] Unsurprisingly perhaps, after these somewhat tortuous negotiations, ratification proved difficult, not helped by the whole package being put to the workforce at each site as a 'take it or leave it' deal.[41]

Although the companies had secured more even wages across the Pilbara, they soon found that the workers were unhappy with the rigidity of that system, not least the more skilled workers. Looking at what had happened in these negotiations more widely, the wage increases were not the most ground-breaking features of these agreements. Rather – in terms of exemplars for workers elsewhere, as well as providing new base lines in the Pilbara – it was the changes to conditions which were innovative. There were pace-setting improvements in leave entitlements (including an extra week each year), earlier access to long service leave, and increases in change-over times and shift and overtime rates.[42] Not all of these union-won changes would stand the test of time, but many did.

In developing a framework for the negotiations in 1976, a novel solution to workplace unrest was to have more, not fewer, agreements, grouping the unions by function and then by site: four groups (two for production, one for engineering and one for building trades) and four companies. This meant there were sixteen agreements altogether. This fragmentation only drove workers' dissatisfaction – and strike levels – to new heights.[43] Robe River wavered and broke ranks to set up an 'attendance and operations payments scheme' for workers, which the financial press derided as an attempt to 'buy peace'.[44] Wage relativities and rationales for them became confused. Little surprise that AMMA's secretary lamented the lack of coordination between the mining companies.[45]

The negotiating parties planned to rework entirely the arrangements next time around, seeking to move from sixteen agreements to just one, but the local unions vetoed this. The ensuing negotiations were marked by reminders that Pilbara realities would always trump the procedural niceties of meetings in Perth. First Robe River and then Mt Newman Mining went their own ways, the latter admittedly after a strike had closed the whole operation. This time, the companies were splintered because the four agreements had different expiry dates.[46]

During this round of negotiations, Mt Newman Mining's workers became more antagonistic not only to the offer put to them, but also to the Commission's attempts to resolve the dispute.[47] For three full weeks in April 1977, the AMWU members were on strike at the Newman mine, with consequent stand-downs shutting the whole operation. This drove the companies and the Commission back to negotiating agreements for each enterprise. When the Newman negotiations picked up again in May, both sides dug in amid growing bitterness all round. The workers rejected the offers; the company, through AMMA, stressed that it would not go beyond the Commission's overall guidelines. Doing what the unions asked – be it matching wages to inflation, conceding more long service leave or introducing the 35-hour week – was 'non-negotiable'.

The AMWU's Denis Day reported to the Commission that 'the attitude of the workers [was] hardening against the company' and that union officials could not persuade Port Hedland and Newman meetings to return to work. Even the charismatic Jack Marks, arguing that the company's viability was under threat as markets changed, could not hose down the strikers' hostility to the company. Commissioner Kelly's decision was handed down in August, the unions having been on strike for much of that time.[48] Workers being threatened with dismissal did not help: there was a 'very rough meeting' in Port Hedland.[49] There were growing divisions within the towns after some had gone back to work earlier than most. Whether a Commission exhortation could protect those branded as 'scabs' in such towns was debatable: 'your type we can do without', wrote the Engine Drivers to one such member.[50] Commissioner Kelly indicated that so defiant had the unions been that he was taking the dramatic step of removing the preference clause from the award. Dramatic it was, but ineffective: unions and company agreed to its continuation.[51]

Across the Pilbara in the 1970s, negotiations through unions for company-based agreements had certainly worked to the advantage of iron ore workers. Average male earnings were almost 70 per cent above the state average, although we must always remember that extensive overtime was being worked and the pay received always included the allowances which the unions had secured as compensation for the harsh physical conditions. Even so, if we compare award rates between the Pilbara and the state overall, the Pilbara's wage rates rose more quickly. In rough terms, allowing for prices, wages rose by about 50 per cent across key classifications. All other allowances and leave entitlements also improved significantly through the decade.[52]

Arcing up...and signs of things to come

The long-standing tensions in the Pilbara came to a head as the industry's markets grew more uncertain in the late 1970s and early 1980s, with demand easing and output barely rising between 1976 and 1979.[53] As global, national and state pressures butted up against local conditions, tensions came to a head. In 1979, there was the Pilbara's longest (and most obviously political) strike at Hamersley, and an experiment with a new management strategy at Goldsworthy; in 1980 redundancies at Robe River created a crisis for the unions; and from 1981, Hamersley's management fought once more to wrest back control from convenors.

The Hamersley strike arose from negotiations to replace the 1977 agreement. From the outset, the unions and the company's managers were united in one thing only: mutual frustration. The strike lasted ten weeks, making clear not only the fundamental tension between workers and employers, but also the complexities of gender and community in the Pilbara and the politics of resources in Western Australia.[54]

As well as the deep-seated tensions in the Pilbara's workplaces, there was a set of specific concerns as the 1977 Hamersley agreement came close to its expiry date of July 1979. In the months leading up to this, the Engine Drivers' Union decided to go its own way, partly because of its members' long-held view that their skills were not being adequately rewarded when they campaigned with other unions. This union's train-driving members had a strong sense of their strategic importance, and other workers routinely referred to them as 'koala bears' – that is, a protected species. As we shall see, despite (or more likely because of) this, the rail section remained one of the most pro-union parts of the Pilbara when times changed.

The rail workers were not the only ones to be disgruntled in 1979. Many workers believed that the 1977 agreement had been a poor one, partly because so much union attention had been focussed on ending the Newman strike which had disrupted negotiations in that year.

For their part, Hamersley's managers were not sitting idly waiting for the unions to hit them up with fresh demands. They were determined to bring the unions under control. An internal document leaked to the press showed that this was a matter of taking on the local convenors and reducing the power of the workforce. Terry Lynch, the company's highly regarded IR manager, set out a strategy which suggested that in dealing with the convenors, ways needed to be found to distract them from strikes 'whilst satisfying their ego' and that increasing utility charges in the towns might 'reduce [employees'] ability to afford to engage in strikes'. Stockpiles of ore should be built up to prepare for a strike. In one area, no expense should be spared – and that was paid travel to, and accommodation in, Perth. One critic had no doubt as to the purpose of this: to 'separate rank and file from delegates during negotiations'.[55] This was a high-stakes game.

The unions' demands were not particularly new. They wanted big wage increases and changes to four major working conditions: how time served was calculated; paid meal breaks; redundancy provisions to accord with Australian Council of Trade Unions (ACTU) policy; and a sickness and accident scheme wholly financed by the company. The novel element in all this – the unions' own high-risk strategy – was to present these changes as non-negotiable. The unions wanted to avoid what they saw as another round in a series of interminable negotiations on similar matters. They may also have been keen to avoid tangling too closely with 'the formidable Hamersley management negotiating team'.[56] This well-resourced and informed team was a specific example of the general power which a transnational corporation could bring to bear in any one site.[57]

Things moved quickly. On 14 May 1979 the unions presented their claims to the company. Eight days later the claims were rejected and the company lodged its counter-claims (itself a novel step). These included the increase in local charges which Lynch had proposed earlier. About 98 per cent of the workforce supported a strike vote. If the workforce was confident, so was the employer. There is some evidence that the managers were quite sure they could easily weather a strike of four to six weeks' duration.[58] They set about attacking the convenors, trying to outflank them by writing letters to union leaders in Perth and to the local workforce asking, in effect, if they really knew what was going on. No doubt they hoped the answer would be 'no'. It was more likely that in Perth the answer would have been 'yes, but we don't know what to do about it', and in the Pilbara 'yes, and we're right behind it'.

Other developments only made the key figures in the unions more determined and arguably more able to carry the fight up to Hamersley. The first was that in all three towns, Tom Price, Paraburdoo and Dampier, families were mobilised in support of the strike in a new configuration of gender roles and community engagement. Women had typically been ignored or taken for granted in previous disputes; but not this time. They set up meetings and funds to support the stoppage. They also established cooperatives and a program to assist families more generally. Coordination and practical support was also shown through film sessions and childcare, newsletters and press releases.[59] The second change was to set up a bank account in each town for the 'Assistance Fund'. Eventually $220,000 (about $1 million in 2017 values) was collected.

The strike was transformed on 11 and 12 June when state and national leaders of the Metal Workers' Union, Jack Marks and

Laurie Carmichael, were arrested in Karratha. This extraordinary decision has caused much debate. Was it really just a local police matter? Did Hamersley know? What was the government's role? But on reflection it is the broader story that really matters. Section 54B of the state's *Police Act*, which was designed to prevent unapproved public meetings, was already highly controversial. A conservative government in another resource state, Queensland, had found it necessary (or perhaps just expedient) to have similar legislation. The tolerance for any kind of protest action was, to put it mildly, minimal in both states. It was into this terrain that the Hamersley strikers and their leaders had inadvertently strayed. The Pilbara was suddenly placed at the centre of state and national debates. On 20 June, two million workers across the country stopped work in protest. As some involved in the strike wrote soon after, it was potentially the case that no democratic organisation such as a union could hold a meeting at all, whereas others could do as they pleased: the 'Directors of Hamersley Iron [were] used to holding meetings behind closed doors'.[60]

If some were uncomfortable with this turn of events, others were not. It allowed the state premier, Charles Court, to run the kind of parochial and reactionary lines of argument that served him and his Queensland equivalent so well: this was all a conspiracy, be it among communists (which, at that stage, both Marks and Carmichael still were), irresponsible unions, or outsiders in general against the good sense and economic interests of the people of the state. The leadership of the police force (if not the frontline men and women themselves) were forceful advocates of the legislation and its use. One can ponder what might have been possible if the political issues had been driven more, not less, strongly by the unions and common cause made with civil libertarians, given what was to be revealed later about miscarriages of justice under the state's hardline policing and judicial practices in that very period.[61]

These wider issues are important because, as one of the radical critiques of the strike suggested, the arrests opened up political questions about life and politics in a resource-dependent state, but it seems that most workers were happy to concentrate on wages and conditions.[62] For all its bleating about workers holding the state to ransom, the government must have been relieved that the ambition of the strikers grew no wider.

In the end, political pressure of another kind was important. After the strike had run for two months, state and federal politicians evidently indicated that they would see Hamersley shut down unless the company brought the unions under control. The company was perhaps not convinced it could do this, or that the governments would carry out their bluff, but union leaders were fearful that at least one of them might at least try these drastic options. Once again, the state's intervention was on behalf of the transnational corporation against the local workers, this time with a threat that the unions would be deregistered. This would mean losing their protections not only in the Pilbara but nationwide. With this pressure on them, the company and unions hastily concluded an agreement. It offered the workers much of what they had demanded, namely their 'non-negotiable' changes to conditions. The wage rise was 20 per cent; half the original claim.[63]

The decision to return to work was by no means straight-forward. When first put to the membership on Friday 24 August, the package was overwhelmingly accepted at Dampier, but comprehensively rejected at both mine sites. A frenetic weekend of persuasion that this was the best possible deal for all of the workforce followed, and on the Monday the workers at Tom Price voted to accept it. No vote was taken at Paraburdoo, and so the deal was declared accepted. Divisions and distrust remained. The more militant workers were aggrieved that the wage increase was inadequate, and that their rents would rise. More importantly, they believed that they had been boxed in by an alliance of the state government, the company and top union officials.[64]

If it was possible to paint the outcome as a victory for the workers – and the other companies followed with broadly similar deals – it was not without cost in all sorts of ways, not least the discontent in the inland sites. It remained an open question as to whether any of the innovative aspects of the strike, such as uniting household and industrial issues, and women and men, or the nascent political questions, would be built upon across the Hamersley sites or the Pilbara more widely.[65]

Independently of all this, the smallest employer, Goldsworthy, adopted a quite different strategy, informed by senior management's reading of German employee participation through that country's statutory system of 'co-determination'. Sitting alongside but separate from collective bargaining structures, co-determination allowed workers a formal say in the workplace decision making. Alfred Kober, the company's GM, believed that poor leadership and miscommunication lay behind the high levels of conflict in the workplace. This analysis may have overlooked the many complex structural factors at play, but it did lead to a specific set of policies to change workplace cultures: in this case a hierarchy of meetings with set responsibilities and with guaranteed worker and union involvement. More generally, the hope was that simply by being in each other's company, workers, supervisors and managers would come to appreciate not only the problems facing the industry, but also each other's solutions and indeed personal merits. Over the next four years, there was a marked decline in recorded stoppages at Goldsworthy, but after that time, when the main pit closed, workers' orientations to the job were shaped by uncertainty about the company's future.[66]

That this kind of approach did not spread more widely is not surprising. Goldsworthy had always been a smaller and somewhat different site to the others, and its towns regarded as more comfortable places to live (and fish) than others.[67] The new process appeared to give everyone something of what they wanted without necessarily bringing fundamental change to workplace relations. Thompson's account suggests that the change was more about information-sharing than genuine power-sharing, and that the local union representatives knew this very well and did take useful data from it while securing some improvements in conditions. For the senior union officials, it meant less work and, in Thompson's assessment, being 'wined and dined' at regular meetings to discuss the program itself.[68] Interviews in the 1990s revealed that many workers were ambivalent: they believed that Kober himself was genuine and that Goldsworthy was a better place to work than the bigger companies, but the ever-present threat of closure played on their minds.[69]

Amid global uncertainty, Robe River's Cape Lambert site saw the first major round of Pilbara redundancies, with the closure of the pellet plant in 1980. For all the high drama and political intrigue of the Hamersley dispute, and the suggestion of a new kind of approach to 'human resources' at Goldsworthy, the plant closure was in some ways more important – revealing weaknesses in the Pilbara unions. For some time, production levels had been falling and, finally, an export agreement was signed under which only sinter ores would be shipped to Japan. The Japanese position should not have come as surprise to the company (and perhaps it did not), nor to the unions given that Hamersley had gone through the same experience only a year before, closing its pellet plant under new contractual arrangements. There were no job losses at Hamersley partly because new facilities to upgrade fine ores were established.[70] Robe River announced that 250 jobs would be lost.

On 11 April, voluntary redundancies were offered if workers registered their interest within four days. This dispute made no sense in purely 'Pilbara terms'. The wider setting was critical: at precisely this time, the AMWU was driving a national campaign to reduce working hours to thirty-five a week. This meant that a closure such as this, in an otherwise strong industry where workers were doing a lot of overtime, provided a perfect site to test these arguments and seek a solution to the closure which involved some lateral thinking.

At first, the combined unions agreed to tie the local problem to the national hours campaign, but lack of commitment quickly undid this. Geography played its part because, for whatever reason, communication between the port at Cape Lambert and the mine site at Pannawonica broke down: the latter had no idea about the state of play, and left the workers at the Cape to their own devices. The Electricians walked away in frustration, while the biggest union, the AWU, changed its position and abandoned the campaign, which created a demarcation dispute with the TWU and left its militant Cape Lambert convenor isolated. The AWU convenor, Dave Dawkins, had provided spirited leadership since 1976, so much so that he had offended union officials 'down south'. When he and his supporters tried to move from the AWU to another union, their erstwhile comrades were happy for them to be the first to be sacked. So much for the workers' solidarity within the Pilbara and between there and Perth.[71]

This messy set of union problems played right into the company's hands. The AMWU was left as the sole champion of the hours strategy, but the position of the local convenor and the stewards quickly became untenable because, in the face of the redundancy offer, most workers were inclined to give up on the hours campaign. According to the closest account of the crisis, 'officials in Perth were…seen as providing insufficient assistance in the face of the backlash from the rank and file'.[72] Local champions of the national strategy were left high and dry. Not for the last time, the most active of Pilbara unionists felt they had been abandoned. The profit drive had triumphed over the union prophets of new working-time arrangements.

Hamersley's approach going into the 1980s was shaped by the managers' reading of the 1979 strike. For Terry Lynch, local workplace processes were 'very bad; very bad indeed'.[73] The continued weakening of the iron ore market only increased his and others' determination to reduce costs. Over the summer of 1980–81, a series of demarcation disputes related to technological changes in the pits only confirmed for managers that the unions were still being wholly unreasonable. On the other hand, the workforce, for all its supposed workplace power, was frustrated at being unable to persuade anyone that Hamersley was already making unilateral decisions affecting work and workers.[74]

For the company, notwithstanding some successes in the Commission, the overall situation required further attacks on the convenors, whose authority in, and knowledge of, the company's operations appeared to be greater than that of lower-level management. From 1981, a strategy was developed to isolate the convenors from the Commission and state union officials, while reducing the convenors' ability to deliver the goods to their members. This was not a matter (directly) of wages and conditions, but of the convenors' power in daily workplace exchanges.[75]

This came to a head in late 1981 when the company wanted to win back control over 'manning', rosters and overtime. In the Commission, Commissioner Collier was sympathetic, criticising the convenors and shop stewards and what he called the 'irresponsible iron ore ethos'.[76] He blamed union leaders for tolerating this state of affairs and the company for having gone along with it. Now it was time for the stewards to be brought under control: 'this mess' had to be fixed.[77] Most of the management's claims were upheld. They could withdraw from previous local arrangements if they wished; unions could no longer intervene in 'manning levels', nor could union rules override the sort of work employees might be required to perform. The unions were told to look at their own procedures: how had the Pilbara developed as it had?

From this case, a new set of formal procedures emerged at Hamersley which delighted the company. Line management's authority was enhanced as was, formally anyway, the authority of the state unions over local workers and convenors. The workforce and its militant leaders had been caught in a pincer movement.[78] The company built on this success, resisting demands for hours reductions in 1983, and securing more control over the allocation of work in return for any gains the unions did make.[79]

Unsettling the Pilbara terrain

Workers in the Pilbara had built what seemed to be a powerful and dynamic collective presence. This worried union officials in the big cities as much as it worried the companies when times got tougher. The Pilbara unions had worked with communities and families to define and control aspects of work and life in the Pilbara. Community and traditional family structures contributed to union power, and at times the gender roles were more complex than they at first seemed – as evidenced by the work of cooperatives and women's support groups.

The Pilbara's unionised workforce entered the early 1980s with material gains and most local workplace structures intact – having survived challenges from employers and the state government. But as the world around them continued to change, it was plain that the workforce's capacity to shape the Pilbara was not always as powerful as it seemed. The companies were beginning to shift the frontier of control, to assert themselves over this terrain. More than one union convenor and 'outside' union sympathiser expressed frustration with the instrumental way in which most workers were coming to regard their own unions. It was all about money, not the progressive role unions could play or the wider politics of the Pilbara and the industry.[80] Strange as it may seem, given how the Pilbara unions are usually understood, these men took the view that, in short, the workforce was not radical enough.

More specifically, through its plant closure, Robe River reacted to an external threat in a way that worked to its advantage – as one account put it, 'before the unions knew what had hit them'.[81] The changes made at Hamersley after 1981 were also instructive: the company had cleverly used the federal and state systems within which it operated to limit local union power. But what might be possible if the companies could find ways to change the overall framework in which employment relations were set? Could they 'refix' the Pilbara more fundamentally?

Part Three:
The Companies Fight Back

In Australia, for nearly a century, most workers' wages and conditions had been set by a system of compulsory arbitration which provided protections to unions and, in fact, could not function without them. These processes and institutions gave legal sanction to the idea of 'a fair go', contributed to relatively high levels of income equality and, from 1972 until the 1990s, even helped reduced the wage gap between men and women.

All this changed from the 1980s. Almost every aspect of the global, national and local dynamics that had shaped the Pilbara since the mid-1960s was transformed. The market for iron ore was less certain than it had been, and competition from other sources increased. Employers were determined to win back greater control of worksites. In this, they were not alone. Around the world, government support for unions, and employers' tolerance of them, had receded. Governments headed by Margaret Thatcher in the UK and Ronald Reagan in the USA tore up the social pacts that had underwritten high levels of employment and wage growth since the end of World War II. With the entire system of global capitalism facing stagflation, falling profits and rising fuel costs, conflicts between unions and employers reached heights not seen since the 1930s.

In the Pilbara, fundamental changes in government policy, law and employer practice came together to pose an elemental challenge to the power workers had created for themselves. The processes unfolded some time after the changes in the USA and the UK, with their own local roots and timings. One by one, between 1986 and 1999, all three major Pilbara operators – Robe River, Hamersley Iron and BHP – set out to de-unionise their operations, trying to 're-fix' the Pilbara.

Chapter Five

War of Attrition: Robe River

FROM THE MIDDLE of 1985, new owners of Robe River's Pilbara operations set out to eradicate union power in a dispute which was reported across the country. It was widely (and rightly) regarded as a test case for how workplaces would be governed as a so-called 'New Right' anti-union strategy unfolded across Australia. The tensions that had been building in the company became public on Thursday 31 July 1986. That morning, many of the senior managers were sacked when they arrived at the company's headquarters on St George's Terrace in Perth. Later that day, the new managers issued a storm of memos about how the company would be run, threatening an entirely new era for the Pilbara's workers and communities. Eleven days later, the managers locked out their entire Pilbara workforce. The lockout, subsequent hearings and then a strike that summer attracted the interest of the national media and aroused hopes and fears in companies and unions beyond the Pilbara.

By late January 1987, the unions had been seriously weakened and, after the strike, things seemed settled. Robe River, and with it the Pilbara, became 'remote' once more as the story faded from the national mind. Employment relations are rarely, however, so cut and dried. The dispute dragged on for six years in a war of attrition until it meshed again with state and federal politics as the workforce finally quit their unions. We need to revisit not only the immediate dispute but also the long aftermath to get a sense of how the Robe River conflict was tied to much wider changes.[1]

Understanding change at Robe River

As the global economy changed and unemployment grew worldwide, the Labor Party, in government since 1983 under Prime Minister Bob Hawke, sought union cooperation in reworking the national economy. The party had come to a Prices and Incomes Accord with the unions' peak body, the Australian Council of Trade Unions (ACTU). Under this agreement, the government pledged itself to improving the 'social wage' (that is, workers' overall standard of living), jobs growth and, 'over time', real wages. With federal unions locked in, the government opened the Australian economy to global capital and drove greater domestic competition. Most unions and some employers, chiefly in the manufacturing sector, embraced this consensus strategy designed to avoid the class confrontations and labour movement defeats that had been a feature of British and American politics.[2] This approach reached into Western Australia, with the establishment of an Iron Ore Industry Consultative Council (IOICC) in 1984. This was a tripartite body of union representatives from the Pilbara, the state Trades and Labour Council, the newly established Mining Unions' Association (a Pilbara body), all the mining companies and the Western Australian and federal governments.[3]

For some conservative think tanks, lobby groups and employer associations, the changes the government was making in seeking to open and make more flexible the Australian economy were too slow and too union-friendly. The newly established Business Council of Australia began to argue the case for what it called 'enterprise-based bargaining units', as opposed to industry or occupational awards.[4] The H. R. Nicholls Society was launched in 1986, its members determined to eradicate union power in wage setting and in the workplace. The title of the Society's first publication, *Arbitration in Contempt*, made plain where its values lay.[5] There was more to this than merely debate. Well before any changes in labour law, some 'New Right' employers went on the offensive. In disputes at the South East Queensland Electricity Board, the Mudginberri abattoir in the Northern Territory, and Dollar Sweets, a small confectionery company in Melbourne, common law torts were used to attack militant unions.[6]

Robe River, or to give it its full name, Cliffs Robe River Iron Associates (CRRIA), was still third in size behind Hamersley Iron and Mt Newman Mining. At this stage, CCRIA employed about 1,300 workers.[7] It was owned by Australian, Japanese and American companies. An Australian company, Robe River Ltd, held the largest stake (35 per cent), but the most important entity was the US-based founding company, Cliffs Western Australian Mining Company, which, despite having only a 30 per cent stake, really ran the Pilbara operations. Cliffs' major owner, with 53 per cent of the shares, was the US parent company, Cleveland Cliffs. It had provided the initial CEO for CRRIA.[8]

Many Robe workers referred to the employer as 'Uncle Cliff', so benign did it seem despite its American ownership and the recent job cuts. The company faced a more difficult environment than others because its ores were lower quality and were sold at 'spot prices', not the more common long-term contracts. With managers more determined than at other sites to maintain continuous production at almost any cost, workers and their unions had held on to their improved conditions. The union convenors exercised control over many aspects of production on the mine and at the port, more so than at the other Pilbara mines. Unsurprisingly, Robe's shop-floor union representatives were sceptical about the ICICC.[9] After two days of briefings about the global iron ore industry, all designed to get workers to moderate their demands, a Robe convenor simply asked, obviously in exasperation: 'When are you bastards going to fix the fridge in our lunchroom?'[10]

In this unstable local, national and global context, the ownership of CRRIA changed. This was the signal that two opposing trends, local union militancy and national consensus politics, might be challenged at Robe River. In 1983, Peko Wallsend, which was a holding company based in New South Wales, bought out Robe River Ltd's 35 per cent stake in CRRIA and, at the beginning of 1986, it acquired the entire holdings of the US parent. Peko now owned a majority share of Cliffs Western Australian Mining Company and its holdings comprised 50.9 per cent of the newly named Robe River Iron Associates.[11]

Peko Wallsend's CEO, Charles Copeman, who was a member of the H. R. Nicholls Society, made it plain that with ownership would come control. It soon became apparent that he intended to bring the unions to heel.[12] Although much commentary at the time argued that the attack on the unions at Robe was an 'Americanisation' of Australian industrial relations, the dispute began after the company was 'Australianised'.[13] Copeman himself was clear about the local roots of change, recording how he 'was inspired by that memorable weekend early in June 1986' when the H. R. Nicholls Society was formed and by members' 'encouragement to initiate' change at Robe River.[14]

The first assault, 1986–87

The new Peko Wallsend management team had one overall goal: restoring managerial prerogative. To sell their strategy to the workforce and the community, Peko's representatives quickly came to focus on 'restrictive work practices', a term which echoed a national discourse developed in response to global economic change. This powerful phrase resonated with the language of New Right employers and lobbyists, and with the emerging agenda of the Accord partners. Globalisation and enhanced product market competition necessitated, it was said, greater 'flexibility' in workplaces. In the middle of 1986, Treasurer Paul Keating observed that unless manufacturing revived and wages were moderated amid 'a sensible economic policy then Australia...[would] just end up being a third rate economy...a banana republic'.[15]

In seeking to shift the frontier of control at Pannawonica's mine and Port Lambert's harbour, Peko faced more than one opponent. The new team had to deal not just with the workers but with the unions who represented them, the state Industrial Relations Commission which recognised union rights, and the existing management which they saw as complicit in the growth of local union power. Copeman's team was experienced in tough industrial settings, no-one more so than his new Industrial Relations Manager, Herb Larratt, who became widely regarded as the author of the statement that 'every worker should go to work each day expecting to be sacked'.[16] Uncle Cliff was to be no more.

According to Copeman, the Robe managers were the immediate problem. He told the H. R. Nicholls Society in 1987 that those managers had been 'at best very reluctant to allow Peko to pursue its own investigations into employment matters'.[17] Three years earlier, on his first foray into Robe, Copeman had been worried about 'over-employment' in Robe, a view he claimed the existing management shared but had done nothing about.[18] Following a demarcation strike in May 1986, Peko sent its own team to the Pilbara 'to investigate thoroughly every aspect of the project'.[19] The Peko executives were outsiders in every sense. They were new to iron ore, to the Pilbara and to Western Australia. They had no truck with any notion of 'a Pilbara way' of doing things and no sympathy for local plans to discuss the company's performance with the unions or the IOICC. The new CEO believed that 'work performances were deteriorating rapidly'. Talking to the local unions would achieve nothing.[20]

Copeman's first presentation to the H. R. Nicholls Society focussed more on the previous management team than the unions. He said that 'site delegates could always override the supervisory staff' but argued that this was only possible because management had abandoned its responsibility to manage.[21] This was a key term: for Copeman, managers had less a right to manage than a responsibility to do so.[22]

The underlying problem, however, as Copeman saw it, was union power. Even a low level of strikes was problematical because this simply meant that 'union demands...were being met without the need to resort to strike action'.[23] Copeman and his team took some time to decide upon their strategy to bring the unions down. When it did come, all the road-blocks were targeted – managers, unions, workers, Commission.[24] On that Thursday, Copeman himself dismissed the doomed managers.[25] A wave of memos and letters then flowed from the Perth office. Ian McRae, the new General Manager Operations, communicated what might be called the Peko way to three distinct audiences. He wrote to the state secretaries of the mining unions, telling them that the company would unilaterally withdraw from award discussions then underway and, perhaps more tellingly, from local agreements governing work on the Robe sites in favour of 'procedures...solely determined by the Company in accordance with sound management practices'.[26] He also told the union convenors in the Pilbara that their privileges were being withdrawn immediately: all 'previous arrangements...are hereby nullified and have no further effect'.[27] Finally, he addressed the workforce itself through a 'notice to all employees'. From now on, he said, 'the employment of any person who permits himself [sic] to be used as an instrument for the imposition of any restrictive work practice is at risk'. In case anyone missed the point he summed it up thus: 'Management decisions will be made by management'.[28]

Communicating with the workforce directly and their unions separately would become the norm later and, under Rio Tinto, would have a greater degree of subtlety, but in the circumstances of the Pilbara in 1986, this was a blunt signal of change. This approach was consistent with the assumptions of those in the H. R. Nicholls Society, and the New Right more broadly, that workers had a more or less natural commitment to the employer and that support for militant union leaders was an aberration.[29] This common enough view misread the peculiarities not only of the industry but of the place. In the Pilbara, the lives of mining workers, families and local union leaders were interwoven in ways which marked the towns and lay beneath the peculiarities of local union power.

The memo flurry of 31 July was the opening of the campaign. One other element remained. For Copeman and his strategists, the Western Australian Industrial Relations Commission was almost as much a problem as were the unions. The struggle in and against the state regulator was mind-bendingly complex. It lasted for seven years. Peko's initial plan was to escape from the state jurisdiction, which had always governed formal employment relations in the Pilbara, in favour of a fresh start with federal regulation and a federal award.[30] Copeman's managers began to go down this path by seeking to frustrate the union and, indeed, the Commission itself. They stalled negotiations with the unions and refused to submit matters to conciliation. Instead, all disputes would be arbitrated.[31] Throughout this time, Peko was highly critical of the way in which the Commission operated. They appeared to rely on the regulator with this focus on arbitration, but in truth their plan was to destroy the power of what they saw as an external body, an obstructive third party.

In response to the new world in which they found themselves, the unions developed a strategy that, above all, sought to avoid strikes or other forms of industrial action. The unions' leadership argued that militancy would play into management's hands. Many workers did urge strike action at Robe itself and across the Pilbara in general. For them, this might draw out their opponents and pre-empt what they believed was a broader attack on unions from the New Right and other employers. Rightly or wrongly, Peko's managers thought that a strike would do their own cause no harm at all. Among other things, Copeman believed that the aftermath of any such strikes 'could provide the opportunity to re-employ selectively'.[32] Even when sixty workers were dismissed for failing to work as directed by management, no strike action was taken. Instead, the unions approached the Commission to have the workers reinstated.

If there was any doubt about the changes in management strategy, it evaporated with the unheard of response to the Commission's order. On 11 August, the management locked out the entire Robe River wages workforce of 1,100 people. This radical departure from local practice not only inflamed the unions, but also deeply worried the state and federal Labor governments, which had been trying to take the heat out of the tensions at Robe.

Behind the scenes, Peko's work practices investigation continued. The management claimed that there were 284 such practices around rosters, classifications, demarcations, wash-up times, union rights and the like. Each threatened efficiency and, more generally, managers' capacity to manage. The mantra of '284 restrictive work practices' was repeated tirelessly in coming months, and taken up without much evaluation by the media, as were claims about strikes over the range of ice-cream flavours in the canteens.[33]

On 3 September, the Commission delivered a compromise to end the lock-out and begin to address the claims about work practices. It ordered the company to reinstate the workers and agreed to investigate the work practices at Robe River. Although the Commission ordered the workers' reinstatement, Copeman was delighted because he believed that he had pushed both the unions and the Commission itself into a corner. He was sure that he was winning the battle to restore managerial power.[34]

On 30 October, after weeks of intense investigation, the Commission handed down its decision on work practices and some other matters. The arguments that the Commission had heard were as much about process as substance. Most importantly, the unions insisted that because these local work practices had Commission endorsement, they were legitimate. They were not some kind of random rort. For its part, the management team argued that it was entitled to make unilateral changes to work at its operations. In a sense, this was the playing out of a story as old as industrialisation – the challenge to the craft workers and the rise of a distinct management function, all around the contested terrain of the workplace. Who would control how work was done?

As to substance, the Commission had been asked by the Robe River management to examine several matters beyond the original 284. The Commission decided that of all the issues before it, there were thirty-two work practices upon which it should concentrate. These matters were of sufficient moment to allow for a manageable way of settling the dispute.

The Commission was highly critical of the management's belligerent approach to change as well as its use of the lock-out and how it had dismissed workers. On work practices, however, it found in favour of the company. Only seven of the thirty-two matters examined were allowed to stand. Of these, an acceptance of paid union meetings was probably the most important, but the others that survived were mostly minor. All the other matters in dispute were left to managerial prerogative. Copeman's confidence that his strategy would succeed appeared to have been justified.[35]

While the arguments were played out in the tribunal, there were more direct and personal conflicts in the Pilbara itself. As managers set about reclaiming control of the worksites through the Commission's rulings, they needed to give effect to the outcomes in the workplace itself. Aggressive confrontations at work became commonplace. The unions were isolated from any effective role in decision-making as supervisors fell into line with the new managerial approach. The more militant shop stewards and workers were isolated from each other with changes to rosters; more dismissals were threatened; militant workers were given 'offers' to leave their jobs that were hard to resist.[36] By the end of October, over 500 workers had walked away from the job, which of course meant leaving homes and town as well.[37] Combined, these changes meant that at least this part of the Pilbara was being 'undone' as a union space.

While many workers supported the no-strike strategy, their uncertainty and resentment grew as they returned to work after the lockout, and as the scope of Peko's determination to reorganise the sites became clear.[38] If it was Australian owners who had driven the change, their methods did seem to come from some of the more notorious of American anti-worker ploys designed to isolate and humiliate key workers. The two most striking – which became central to union and family memories of the dispute – were the setting up of special work groups. One was the 'Special Projects Crew', which unionists renamed the 'A-team'. Its members were given tedious, if not wholly unnecessary, jobs to do. The other was the 'Clean Up Group', quickly christened the 'grot squad', whose members were assigned menial tasks off the worksite and in the more public space of the town.

These sorts of jobs, the unions claimed, had once been carried out by prisoners.[39] The humiliation sent a strong message to workers and family as to the consequences of stepping out of line. As the towns became more divided between wage-hands and salaried staff, some workers' children endured, reported one union source, 'the jeers of classmates whose scab parents had become "white hat" supervisors'.[40]

Local workers' patience with their unions' defensive strategies came to an end in December with a 'shovel manning' dispute on the mine at Pannawonica. Familiar enough as a general kind of resistance to the logic of the managerial control of work, this conflict emerged when the management sought to have one worker operate what had long been a two-person power shovel. An initial work stoppage took place on 9 December, followed by a complete shutdown of Robe's operations from 16 December.

This dispute generated still greater division between the company and the workforce and also between local workers and state union leaders.[41] As the workplaces and the towns divided more sharply between unionists and 'white hats', local workers themselves became more divided from one another. An anti-union group of 'concerned workers' set up a network and distributed a news-sheet attacking the strikers and the unions more generally. The unions struck back with a newsletter called *Pekobusters*, *Resist to Exist* sheets and the occasional – and often very funny – *Scabline*, a response to the company's *Jobline* sheet. Friendships and families fell apart as some workers quit their unions to side with Peko. Intimidation on the job from supervisors was widely reported at union meetings and in union publications. The media, predictably enough, preferred to focus on claims of violence by union members.[42]

In January 1987, federal union officials led by ACTU President Simon Crean met Copeman and negotiated a proposal to resolve the shovel dispute. On the ground, however, frustration was not diminishing.[43] A journalist's account nicely captured the local context when Crean arrived in the towns to sell the deal:

> When he landed…at 8.30am…the industrial relations climate was baked as hard and as hot as the red earth and the straggling spinifex. Simon Crean quickly took off his coat and tie. He was about to learn the hard way of the distance between the Pilbara and the rest of Australia.[44]

On 16 January, workers' mass meetings at Wickham and Pannawonica rejected the Crean–Copeman package.

This was to prove, perhaps surprisingly, the high point of defiance. Soon 'the rest of Australia' did reach into the Pilbara. With that increased national influence came the decisive steps in defeat for the local unions, although that was not clear for some time. The immediate causes of the end of the shovel-manning dispute were the company's pursuit of writs for damages against individuals and the unions, including the maritime unions and individual officers. The maritime unions were particularly exposed to legal threats in the federal tribunals and were ordered by the Commission to return to work.[45] In other words, federal judicial power did begin to reshape the Pilbara.

Discussion continued behind the scenes and, just eight days after the rejection of the settlement, there was another set of meetings. This time, on 24 January, after long and angry meetings at Wickham and Pannawonica, the strikers voted to return to work. In doing so, they accepted an agreement which was little different from the one they had rejected a week earlier.[46]

In her account, Pam Swain argued that the defeat of the first Crean-Copeman package was unsurprising because 'the causes of the conflict were emotional and related significantly to loss of power and status by the convenors and shop stewards. The issue could not be addressed by rational arguments put by outside authorities'.[47] There is certainly some truth in this, because it hints that geography, in terms of the location of power and sense of place, lay at the heart of the Robe River struggle. To overplay the emotions beneath the strike and its aftermath is also to suggest that 'outside authorities' were necessarily more 'rational'. That is another thing altogether. Each side had its own logic behind its actions. What is more important is that, armed with judicial power, those outsiders were becoming more powerful.[48] With that change, the managers would also come to exercise more power at work.

A war of attrition, 1987–93

For many observers, the return to work in late January meant that the Robe River dispute was over, with merely a logistical exercise remaining to finalise the investigation of work practices. Since then, most accounts of the dispute have read its timeframe in just that way. The dispute was also understood in a particular geographical way. The Pilbara's remoteness made it easy to imagine the place as wholly unique. So, after January 1987, Robe River certainly faded from media, popular and political attention. This reading of time and place completely misses the long aftermath of sustained management hostility to unions, a beleaguered local resistance and the national implications of what had happened there.

Over the next six years, a series of incidents both large and small (often seemingly trivial) revealed the determination of the Robe River management to remake its workplaces entirely. A war of attrition developed despite the fact that the agents of change, Charles Copeman and Peko Wallsend, had quit Robe River (a fact not much remarked upon). Within a few months of the end of the shovel dispute, North Broken Hill, one of the oldest Australian mining companies, began takeover moves, and then in early 1988 merged with Peko Wallsend.[49] North Broken Hill had a long history of good relations with unions, as part of the Collins House Group of companies which had long negotiated collective agreements with the Broken Hill unions – agreements which had underpinned over sixty years of industrial peace.[50] To Copeman's delight, however, the company's strategy and much of its management personnel were left unchanged.[51] He was able to tell the H. R. Nicholls Society that things 'would not be different' after Peko.[52] It was a sign of things to come that the recent history of Robe River overrode the practices of the Pilbara and the longer traditions of Broken Hill.

As management consolidated its position, arguments about how the dispute had been waged intensified.[53] The key problem for the workers and their most militant leaders was that they faced two forces of change at once: the politics of consensus and the New Right's managerialism. The more militant section of the Robe workforce favoured an all-out counter-attack on employers. This was the 'POP' strategy ('pull out the Pilbara'), much derided by most union leaders.[54] It was premised on the view that, even if some of the practices at Robe were hard to defend, other Pilbara employers would follow Robe's example, as would employers in other places, unless there was fierce resistance at Robe. Jack Marks was later quoted as saying that many of the work practices were 'pretty dodgy'. 'My position was to fight against these stupid ultra-Left, mindless, militant actions which would have led us into an ambush where we would be destroyed'. Prime Minister Hawke and state Labor

Party leaders evidently shared this view, if not necessarily in the same terms.[55]

For their part, local leaders and the union loyalists felt betrayed. While others might have seen the Pilbara as unique and isolated, it was not quite like that for those who were living there at the heart of this dispute. These workers could not be persuaded of the merits of the Accord and consensus, especially when being belted by their employer. One worker reflected that he and others saw their unions not as isolated or unusual but thoroughly tied into working life across the country. They were, he said, 'setting standards for all of Australia'.[56] For them, the mining town of Pannawonica was the key. It would 'break through' with improved wages and conditions and others would follow.[57] This could not happen without a strong local union network and 'worker control of the union' and, with that, of the job.[58] It was precisely these networks and this control that Copeman had taken on and which his successors pursued thereafter.

Unions were not 'outlawed' — to do so was neither legal nor even possible — but a block to union power could be achieved in effect. Unions' capacities as active agents genuinely able to represent workers could be all but eradicated. This is what happened. After 1987, there were almost endless disputes about the general right to hold union meetings and about the specifics of when and where the meetings should be held. In March 1988, the management of Robe did something unheard of by insisting that workers sign a 'personal commitment form' to receive the wage rise from that month's National Wage Case.[59] A little over two years later, at the end of 1990, the company refused at first to pass on wage increases sanctioned by the federal tribunal under the 'Structural Efficiency Principles'.[60] Employment numbers continued to fall; crewing levels were reduced, including in the rail section where the unions had been so strong. Even some managers at other sites had reservations about the wisdom of these moves in terms of safety and production.[61]

The company also unveiled legal and corporate strategies against unions which were soon copied elsewhere. Although the work practices case and the shovel dispute had undercut the mining unions, the tugboat officers and crews remained well-organised in militant unions. One vital part of the production chain was, therefore, still in the hands of unionised workers. Their agreements allowed for some workers' control, anathema to the New Right, over hiring, through 'preference for unionists' clauses. Robe set up a new company to employ the tug crews in September 1990, as Patrick Stevedores would also attempt, aiming to de-unionise its wharf operations in 1998. Determined to eradicate the last union foothold, the employer took this matter to the High Court in 1991.[62] In the 2010s, disputes over the tugs were still a feature of work in the Pilbara.

Then, in 1992, Robe dismissed members of a union new to iron ore mining. The Construction, Mining and Energy Union, as it was then known in Western Australia, had entered the Pilbara when the Engine Drivers' Union joined the federal construction workers' union, which in turn merged with the coalminers' union as part of the ACTU's drive to reduce the number of unions for more efficient performance, if not survival itself. The workers had joined an ACTU national day of action over the Victorian government's new anti-union policies. In some cases, the workers were told by the company they must sign a no-strike pledge to get back to work. Those who refused to do so, twenty-nine in all, were dismissed. When the ACTU's new President, Martin Ferguson, urged the workers to concede, the resentment over the 1986–87 disputes resurfaced. There was a local strike to protest against the company's attitude and, it might be guessed, the ACTU's position. Feelings intensified when workers were threatened with eviction from their homes. For workers in the Pilbara, this would have all-but-guaranteed them losing their jobs and having to leave town.[63]

Even this list of specific interventions by the management of Robe River understates the extent of their determination. In the towns and on the sites, anti-unionism was being played out in what the union workers claimed was constant intimidation. In another new feature of working life in the Pilbara, the management took civil action against workers who went on strike, a precursor to anti-union ploys that would unfold at Hamersley Iron under Rio Tinto from 1992.[64] Each of these episodes eroded what little remained of union power, building on the Copeman success of 1986. Nonetheless, Robe River's managers still had to deal with the Commission and the unions in almost all of these disputes. On one estimate, membership was 60 per cent of the workforce in 1989. The un-named official reporting this figure saw it in terms of the decline in membership. It is striking that it had remained so high, given management hostility and the unions' declining capacity to do much for their members, with the company applying the narrowest interpretations of the award.[65]

At this stage, the Perth-based Labour Council and the mining unions were still funding a last-ditch attempt to hold on to the Robe workforce. In the early 1990s, this attempt collapsed amid all-too-familiar inter-union tensions locally, and against the backdrop of national leaders trying to resolve a carve-up of coverage between the two key unions – the AWU and the CMEU. Dave McLane, working by the early-90s for the CMEU, recalled that local union representatives cooperated pretty well, but the fear induced by Peko and the constant and wearing legal battles wore down all but the most committed unionists.[66]

Nonetheless, there were still obstacles to managerial control of the workplace. One option still pursued by managers was federal coverage of Robe's operations; it was hoped that a federal tribunal might have a different approach from the state's. Another, which was more practicable in the short term, was to obstruct any initiatives for change, from either the unions or the Commission itself, within the Western Australian jurisdiction. So, when the Commission proposed a new, stripped-back agreement, the company rejected it. In it, there was a provision for 'site representatives' in dispute resolution. No doubt this alarmed those who had spent a lot of time and money in eradicating collective representation.[67] For their part, the beleaguered unions persisted in trying to bring the company's negotiators back to the table. Defying the company, the Commission made a new award to bind Robe River in November 1990.[68] Minimal as its role now was, the award system still stood in the way of total managerial control.

The final step came after the state election of February 1993. A conservative government was elected, part of a string of such results across the country, despite the Labor Party retaining federal office. The new Western Australian government, under Premier Richard Court, was keen to do as Jeff Kennett's government had done in Victoria, all but eradicating arbitration and the award system.[69] These governments were falling into line with the new anti-union consensus that had emerged a decade earlier in the UK and the USA.

The new labour laws in Western Australia did not go quite as far as Victoria's, but they restricted arbitration and union rights and promoted individual contracts, Western Australian Workplace Agreements (WPAs). Meantime, on 23 August, Robe finally secured the cancellation of the Robe River award in the Commission. Its solution now was not, though, the federal arena. On that very day, according to the *Australian Financial Review*, Robe 'moved…to set up workplace agreements in line with legislation which has yet to pass through State parliament'.[70] Unions tried to represent their members, and at least change the terms in the new WPAs – but on one account this was almost 'farcical', not least because (anticipating federal law from 1996) the contracts were not individual at all, but template agreements.[71] The point was to exercise managerial control unencumbered by unions or the Commission.

The state laws, which had afforded some protection to Robe's workers, were now working against them. The Pilbara landscape was changing too. By now, Hamersley Iron was using common law torts and these new laws against its unionists. As the next chapter shows, this seemed to happen almost overnight, unlike the long struggle at Robe River.[72]

Increasingly besieged, isolated, and with almost all their experienced local leaders gone, the Robe workers accepted these new individual, non-union agreements which were sweetened with wage rises. Ironically, the very failure of the original Peko Wallsend plan to take the company into the federal jurisdiction now made it easy for the new owners to introduce individual agreements. The company's anti-union energy was undiminished. In December 1996, with 97 per cent of the workforce under WPAs, Robe was back in the Commission arguing against the continuation of an award payment for sickness and accident insurance for just five members of the AWU.[73]

Robe's tactics became a blueprint for other employers, merging with somewhat more sophisticated strategies, as we shall see, adopted by Rio Tinto across New Zealand and Australia. Former Peko Wallsend executives were central to major disputes, such as the APPM dispute in Burnie in Tasmania.[74] The H. R. Nicholls Society lobbied for change at the federal level as many of its members became more influential in national politics. This goal was realised when the Howard government rewrote labour law with the *Workplace Relations Act* in 1996.

In 2000, Rio Tinto, by then the resource sector's leading advocate of non-union employment, purchased a controlling share of the Robe River operations. This time, the work of de-unionising had already been done. Its other Pilbara arm was the company long regarded as the jewel in the Pilbara crown, Hamersley Iron. How that company became non-union is the story of the next chapter.

The fall of a union stronghold

In one of the few contemporary accounts of the initial dispute, Thompson and a co-author, Howard Smith, well understood where it might lead. They located the Pilbara in a national setting:

'Whether or not the views of the so-called New Right will become a permanent component in the industrial relations debate depends to a large degree on the consequences of the dispute at Robe River'.[75]

The local impact was comprehensive. It may have taken political change at the state level in 1993 for the Robe managers to complete the rout, but the ground had long been laid for what was once unthinkable: that the workforce would walk away from its unions. A media statement released as early as 14 August 1986 by the Confederation of Western Australian Industry jumped the gun, but it was a more accurate summary of what was at stake than some other judgements: the dispute 'had shown the great Pilbara experiment in worker participation and industrial democracy to be a failure'.[76] This hostile comment summed up what the workers and their unions had been trying to achieve, if in incomplete and sometimes unclear ways, since 1966. After the Peko assault, the remaining union loyalists were isolated and, they felt, betrayed by their union leaders.

By 1993, the issues that had seemed specific to the Pilbara or a small group of New Right employers had become much more common across the country. The Peko Wallsend strategy was, as some had feared and others hoped, one which heralded change beyond the Pilbara. As Thompson and Smith, along with others at the time, had suggested, though, this wider impact was not simply after but precisely because of the defeat of the unions over 1986–87. Portraying these workers as out of step and out of place with the global and flexible economy was important and enduring. Working lives are still marked by these events. Even by 1993, the world of work and employment relations was very different from seven years earlier. Nationally, union membership was down from 46 to 38 per cent of the workforce, almost every state government in Australia was turning its back on the traditional arbitration system, and the Accord was on its last legs. New threats to traditional ways of working were emerging – and the Pilbara was at the heart of them, as we will see in the next chapter.

Chapter Six

Frontal Assault: Hamersley Iron

As THE LONG struggle to de-unionise Robe River came to an end, a much shorter and sharper conflict was being played out in the biggest and generally most profitable of the Pilbara operators, Hamersley Iron. In June 1992, the workforce appeared to be united behind its union leadership in a dispute central to the Pilbara way of doing things – a strike against the employment of a non-unionist. Yet just eighteen months later, in December 1993, almost all the workers walked away from their unions to sign individual non-union agreements.

This was a transformation of the Pilbara, one which took almost everyone by surprise, a seemingly inexplicable collapse of a union fortress. In order to understand what happened, we need to go back to changes that were underway in the parent company from the 1970s and examine the de-unionisation of another site in the Rio Tinto empire, in the south of New Zealand, before coming back to the immediate backdrop in the Pilbara. The Robe River dispute had attracted much publicity (and rightly so) as a fundamental shift in Australian practices, but the legacy of the Hamersley Iron de-unionisation was at least as widespread and enduring. Rio Tinto developed, preached and won converts to its message and mission of non-unionism or, as it put it, 'direct engagement' with workforces.

Rethinking employment relations in CRA

The origins of direct engagement and, with that, the changes at Hamersley in 1992–93 lie in the 1970s when Roderick Carnegie became both CEO and Chairman of CRA, the forerunner of Rio Tinto.[1] He was generally respected across the company and across political lines as an old-style manager who knew mining from the bottom up. He had degrees in physics and agricultural economics. As a student he had had close contact with L. B. Robinson, who was highly experienced in Broken Hill's unionised mines. Carnegie was one of those managers who was smart enough to know there were things he did not know, in this case much about managing employees. He listened to Robinson and then, more than twenty years later, after taking over CRA, he travelled widely outside Australia seeking his own answers to the problem of managing work in the Pilbara.

He recalled that at this time (1978) he 'read an unreadable book' which included a diagram that made sense to him as a former physics student.[2] It appeared to be a scientific approach to management, a specific way to achieve the often discussed 'shared values' at work. He began to push the idea that the company should take actually seriously the cliché that 'people mattered' more than anything else. For him, 'each person' should be encouraged 'to give of his or her best'.[3] Critically, workers had to know they would be treated fairly by the company or any hope of better workplace relations would crumble.[4] Genuine leadership was, said Carnegie, 'vital'. Leaders had to be 'accountable', doing more than simply managing. To overcome workers' anxieties and suspicions, job security was an essential underpinning of this approach.[5]

The author of the apparently unreadable book was Elliott Jaques, an original multi-disciplinary theorist at home in fields of thought ranging from psychology to politics. Unusually for a management thinker, Jaques was not afraid of writers who took different paths. Few management theorists have quoted from Karl Marx and Friedrich Engels as approvingly as Jaques did, noting that working was about more than money because it was a social process through which men and women defined themselves. Whereas Marx and Engels, along with generations of political activists and workers at the 'frontier of control', argued for ways to seize command of work, if not society itself, for those who produced its value, Jaques turned this argument on its head. He said that communists and other radicals had understood the problem but not the solution. The best way for workers as human beings to realise their potential was not to limit the market and the rule of their managers, but accept and work within it all.[6] It would be necessary therefore to develop 'shared values', as consultants such as McKinsey (where Carnegie had worked) were arguing by the early 1960s.[7]

According to Bruce Hearn Mackinnon and others writing about Rio Tinto, there was nothing unusual about CRA's managers exploring ideas such as these. The company was more innovative and open to global intellectual influences than most, not least more open than the conservative and bureaucratic BHP. Eventually, this mindset and the influence of Jaques led to a more comprehensive and thought-out change program than the simpler anti-union strategy which Peko Wallsend had deployed in Robe River. Carnegie recruited Jaques to CRA and devoted unique levels of 'energy to developing and implementing…an un-chartered theory'.[8]

At first, there was nothing in Carnegie's language or philosophy that was obviously anti-union. There was, though, a challenge to the specifics of the Pilbara variety of unionism. Line managers had to have 'the right of veto' over staffing and work allocation, as well as sole responsibility for individual performance reviews.[9] If managers were to be accountable, then that meant they must have more authority.

Jaques' ideas were labelled 'stratified systems theory'. This meant introducing flatter management structures and being apparently more 'scientific' in how work was organised and roles assigned, with workers' abilities tied to particular jobs. To achieve this, companies were thought of as having five strata, each defined by the time span of a job, ranging from a few weeks for operators to five to ten years for a managing director. While the high-profile changes and fights were underway across the Robe River sites, change was more quietly underway inside CRA's Hamersley operations along the lines that Jaques proposed. There was a significant delayering of management as flatter structures were introduced. Remaining managers were retrained.

When Jaques thought about collective bargaining and unions, he was doubtful about their merits. In fact, even individual bargaining was a problem because any kind of bargaining was about power and relativities; it was not really 'rational'.[10] Jaques' plans, therefore, gave much greater power to managers over work, annual reviews and salary. This question of managerial prerogative was one of the areas that would lead to arguments about the say the workers collectively, notably through their unions, would have in the workplace.

In 1986, Carnegie left the company and it appears that Jaques' influence declined relative to one of his former students, Ian Macdonald, a university psychologist who took a role as a consultant to CRA. The focus of planning moved more from Jaques' stratified systems theory to a plan for 'all staff employment', which posed a more explicit challenge to unions.

Despite continued assertions to the contrary by CRA and then Rio Tinto, the strategy was becoming more obviously anti-union. Some of the leading thinkers in the company were arguing that it was not possible for workers to be committed to both the company and their union. To create a new team environment, CRA's managers had to offer a positive alternative to union representation. Partly because of Carnegie's legacy, the company worked to make the all-staff operation an attractive option. CRA would also make it hard for workers to stick with their unions, mainly by stalling on negotiations for collective agreements. The company was to use the carrot and stick to change work and employment relations in Hamersley.

The first public sign of what all this planning might mean appeared far away at Tiwai Point in 1991. This was the site of CRA's New Zealand Aluminium Smelter, at the southernmost end of the South Island. The smelter was surely as physically isolated a major worksite as could be imagined, yet, like the Pilbara itself, it was cast into the epicentre of national and international change. In a campaign five years in the making, but only out in the open in 1990 and 1991, CRA took advantage of changed national labour laws (the *Employment Contracts Act* of 1991) and of worker and union divisions to de-unionise a site which had been a tough union heartland.[11] As they began to put theory into practice, isolating the unions was a vital part of the plan. CRA was now hardening up: new managers came in who were less knowledgeable about the subtleties of the theory, and more interested in cost reduction. To achieve this, they wanted greater control over the workplace. Workers who were combined in a union were an obstacle to both aims.

The tactics used to get to the endpoint of 'all staff' arrangements became the template for the company across Australasia. At Tiwai Point, almost all the hallmarks of the new strategy of all-staff employment were rolled out. This involved first bringing in new managers or retraining old ones, and then making minor changes to internal procedures to enhance managerial control. To isolate the unions, there was direct communication with the workforce, and there were generous redundancy packages for union convenors and 'known militants'. More generally, the company offered increased salaries without changing the union agreement and, in fact, stalling in those collective discussions. There were also threats of, or actual, downsizing (as opposed to Carnegie's original vision of job security).

The man who had started this strategy had reservations about going to war with the unions. Carnegie later wrote that 'the same results could have been achieved 'over about a five-year period, and without all the associated pain and controversy'.[12] That rethinking lay in the future. In the meantime, a CRA team wrote up a report of what had happened, from which managers could learn. Australian sites were next, including CRA's prized Pilbara operation, Hamersley Iron. 'All staff' employment was coming from Tiwai Point to Tom Price, Paraburdoo and Dampier. As fundamentally important as the Robe River dispute was, it would have been wise for unions to have looked across the Tasman to see what CRA had in mind.

Local struggles: Hamersley and Mt Newman

While all this was happening in CRA, things were already changing back in the Pilbara. There had been significant changes in Hamersley after the company's success in controlling the unions in the early 1980s. New awards were made that pleased the managers, while a massive dispute convulsed Mt Newman's operations.

Changes at Hamersley were driven by management anxiety over the state of the market for iron ore, and were facilitated by national policy responses to a more generalised crisis in profit levels and global competitiveness. The same mantra around flexibility which had surrounded the work-practice disputes at Robe River drove the Accord partners, the Labor government and the ACTU leadership to use the award system to drive workplace change. According to Swain's sympathetic assessment of internal company documents, Hamersley's management team was ahead of the federal government's agenda, keen to 'exercise its prerogative duty in the conduct of the business' in securing 'greater flexibility'.[13]

Company proposals to restructure job classifications and guarantee continuous supply met a cool response from the unions when first discussed in October 1986, but by March 1987 the unions appeared to be much more accepting of these ideas. Swain suggests that the managers attributed this to national developments. National wage polices had introduced a two-tier wage system under which real wage increases were only possible where unions agreed to discussion about workplace flexibilities. No less important, the managers said, were local circumstances: 'the changed climate' in the Pilbara after the initial Robe River onslaught.[14] Unions had slipped into a defensive mindset, at least for the moment. Beneath these immediate issues lay the company's earlier success in controlling the local union convenors and stewards in 1981. Unlike the Peko managers at Robe River, they had not alienated the state Commission but had accepted its role and used it to contain the unions – and, in one more Machiavellian view, worked through the Commission to bureaucratise workplace control and, in so doing, pull the unions' teeth.[15]

The 1987 Hamersley award in some ways put the Pilbara ahead of the rest of the country – this time with management taking the lead. A long preamble to the new award set out the rationale for change. It explained that unions and the company had agreed to more flexibility and that, while the workplace climate had been altered since 1981, there was still 'further scope for improvement'.[16] The number of job classifications was reduced by almost half and there was a greater emphasis on competencies in defining each level. The traditional link between job 'ownership' and union coverage was broken. The award's clause 32, which set out new wage rates, dealt explicitly and at length with the end of the old demarcation practices. The swing in power to supervisors was quite clear. An account sympathetic to management summed it up well: 'any member of a work group [could] be requested to do a job...within their competence'.[17] When interviewed a little over three years later, one of the company's most experienced Pilbara-based managers, Bob Dickson, was quite clear about what had happened:

The '87 award's not bad for the wages workforce, but it's perceived to be bad by giving the company more power, and I think that's a true realisation. That's a fact. The company did get a lot of power out of the '87 award.[18]

He went on to say that the company had continued to drive change in the two to three years after the Award, tolerating participation but insisting on managerial control.[19]

Despite having negotiated these changes with unions through the existing arbitration system, the company was not at all persuaded that negotiation with unions was the way of the future. Swain's account makes this clear. She notes that in 1984 the company's internal polices had explicitly stated that it had 'to be recognised that…notwithstanding the most altruistic of managements…it is part of the nature of employees in Australian industry to see themselves in social conflict with the employer'. Just four years later, with the new award in place, the new version of this policy stated that the goal lay 'in establishing rapport and identity of purpose and commitment between the company and its employees'.[20] To this we might add that the shift in thinking plainly showed the influence of Jaques and signalled that further changes in the relationship with the unionised workforce were, at least in management's mind, inevitable.

The management mindset was obviously changing – and continued to do so because in just four years Hamersley was to be the local focus of this global corporation's de-unionisation strategy. Before turning to that, however, we need to move across the Pilbara to the worksites and towns of Port Hedland and Newman where, in 1988, the social conflict to which Hamersley's managers had referred was playing out in the most vigorous of ways.

Just as new ownership and control at Robe River had been the catalyst for change there, so it was at Mt Newman Mining Company after BHP stepped in. BHP spent $880 million (over $2 billion in 2017 values) to increase its stake from 30 per cent to 85 per cent, effective from 3 January 1986. Finally, the biggest of the old Australian mining companies was running its own show in the Pilbara. Or was it? How would the workers and their unions respond? It was immediately clear that BHP considered that the operation was running far below its potential. The joint-venture structure had been one problem, but now BHP's focus could shift to the workplace. The tradition of buying peace was to change. Further, and many said at long last, local managers would have more say than their bosses in Melbourne.[21]

Like Hamersley, the company drew on McKinsey for consulting advice, seeking to change internal structures to enhance 'responsibility and accountability'.[22] Unlike Hamersley, the company was riven by tensions between levels of management. It could not resolve these, nor work out a viable plan to deal with the unions. Falling between the belligerence of Robe and the tactical acuity of Hamersley, the BHP team rolled from one crisis to another. Within eighteen months of developing a 'work improvement program' in May 1987, the sites were torn apart in the most extensive and bitter series of work stoppages that Newman and Port Hedland had ever undergone.[23]

The senior managers' immediate concerns were driven by cost structures and by what they saw as a flat market and wasted opportunities even in that context. They believed that a new marketing strategy was needed to boost revenue. This analysis was no doubt driven by BHP's desire to recoup some of its massive outlays sooner rather than later. In turn, of course, this meant increasing production and reliability of supply. How exactly that could happen while reducing costs remained to be seen.

The first step in addressing these challenges was to undertake a staff survey, the results of which revealed high levels of discontent over leadership and what was termed 'industrial relations'. This appeared to mean, at least in part, that workers were unclear about who was responsible for what. Therein lay a major problem, which became even clearer in 1988 in interviews with managers and supervisors. It is only a mild exaggeration to say that neither understood each other's work and that each layer blamed the other for the company's problems. The Commissioner dealing most closely with the company, Jack Gregor, was well aware of this. He lamented that in negotiations between unions and 'industrial relations professionals', there was always 'a third party not actually present', namely, senior management. The managers who were present during these negotiations referred to the company as a separate entity from them, which made crystal clear to the Commissioner that there were 'separate management groups within Newman'.[24]

Members of the senior management team believed (rightly or wrongly) that the union convenors were more able and better informed than their immediate supervisors, the lower levels of management. Apart from being an interesting signal that managers (as opposed to media coverage) acknowledged that the power of the convenors was not all bluff and bluster, this was a telling comment about the tensions inside the managerial structure in the company. When the new managers were developing their initial change proposals, there was no indication that the power of line managers over workers would be increased. In fact, many of their own jobs were to go. So thoroughly divided was the management of BHP's Pilbara iron ore sites that many supervisors went on to side with the workforce and the unions against the company's agenda.

With these structural problems to address and with market problems to be faced, the new CEO, Gordon Freeman, declared on 30 June 1988 that the operation was in crisis. Although the company had had equipment failures and suffered from weather problems and train derailments, he now believed that 'bans imposed and a multitude of short term stoppages' were the major problem.[25] It is easy to see that the workers might have been sceptical about this interpretation: they knew that productivity had risen, and employment had fallen to just under 3,500 in 1988, while automation, new equipment and a shift to ore trains of 240 cars (up from 180) had all been introduced. Surely things could not be that bad.

The workers, therefore, doubted the depth of the crisis and were sceptical about the viability of the growth strategy. To many of them it looked either foolish or contrived – they were unimpressed with the new team and sure that reducing worker voice was the real agenda. Nonetheless, senior managers were firm: productivity and work practices had to change – and within a month. The key was to secure a greater return on capital through running equipment for more time each day and spending more effort on the most productive parts of the job, especially in the pit. The prospects for the ordered change that management wanted were, however, slim. Small-scale disputes and stoppages were already breaking out at the Newman site. These were only halted when, in July, the Commission got both sides to agree to a moratorium: no enforced changes and no work stoppages for two months.

That working life in the Pilbara was always complex immediately became clear – a stiff reminder to the new owners. A dispute erupted over a supposedly non-work issue, housing policy. With most parties supporting a buy-back scheme for homes that workers had been renting (part of the normalisation process), a Port Hedland manager sought to sidestep the one remaining sceptic, the AWU's convenor at that site, by going over him to the members. A stopwork to protest against this led the company to seek to change the formal 'industrial relations procedures', approved by the Commission, which governed its relations with the site unions. This stoppage broke the moratorium and brought to a head the long-brewing tensions.

The company's leadership now decided to break with the past: convenors had to be brought under control by supervisors applying the letter of the law. Convenors had to go back to their non-union work tasks. Interestingly, the national processes which some saw as weakening shopfloor unions were read by CEO Freeman very differently. He believed that those reforms which focussed on workplace change and jobs restructuring might make workers believe that they had the right to more, not less, say at work. Wildcat stoppages were now breaking out on different parts of the sites as new deadlines pressed in on management. Two senior mine managers 'resigned' in this increasingly tense environment.

The unions were willing to offer continuity of ore supply if the company agreed not to push for change to the convenors' powers within existing agreements. However, when two convenors were dismissed in late October, the members of three unions went on strike. Electricians and AMWU members remained at work partly because of inter-union tensions over pay relativities. But when management wrote to staff (that is, to white-collar workers), telling them to do wage-hands' work if required, the unions united with each other. Many supervisors sided with them. Some supervisors were sacked; dismissals were also threatened for several hundred workers in the pit at Newman. Divisions tore through the towns, especially Newman. Not only was there mutual abuse but violence too. This gave the lie to the notion of the Pilbara as a homogenous community, but of course it had always been a fiction. As we have seen, defining the meaning of work and life in the Pilbara had come to be cast in class terms. [26] Never was this plainer than in Newman in 1988.

On 7 November, the parties were back in the Commission seeking to resolve the disputes over the use of staff labour and the status of the convenors. The company was now as insistent as Peko Wallsend had been at Robe: the convenors were out of control and the change process was far too slow, especially when the company was facing external market problems and internal cost pressures. The unions maintained that management's divisions were the core problem. With so much ill-will, the Commissioner decided negotiations were all but pointless: he varied but did not cancel the formal procedures. As to other matters in dispute: staff members could get their jobs back. The unions had emerged quite well from this, but were not wholly satisfied and wanted to stay out on strike, which they did until 22 November. In the meantime, the towns remained violent and divided; some managers resigned or were sacked; Hamersley Iron had picked up market share.

In the end, the convenors more or less survived intact and the unions regrouped. In fact, they grew: so distrustful had supervisors become that they had joined the union which covered their work, the Association of Draughting, Supervisory and Technical Employees. Many other supervisors had simply quit.[27]

The Beales dispute and Hamersley's de-unionisation

The strikes and other stoppages at BHP in 1988 tended to obscure the fact that throughout the 1980s there was a marked decline in overall levels of recorded industrial conflict in the Pilbara. In particular, Hamersley had been, if not quite strike-free, then close to it after the 1987 award. The company had been quietly making changes. Employment numbers had come down from about 4,000 at the start of the decade to just over 3,000 by its end. Production had risen, which indicated that so too had productivity. Many of these developments were in line with how the managers read Jaques' stratified systems theory, but they cannot have done much for workers' feelings of job security.[28]

Despite the good numbers to which the company could point, senior managers were less than satisfied. The environment they faced remained turbulent: globally, prices were unstable; competition in Japanese markets was growing from Brazil's massive state-owned firm, Vale; locally, Robe had reduced its workforce size while de-unionising; BHP, as we have just seen, had been in chaos, unable to rein in its unions. Nationally, the government and the unions were still talking consensus as the way forward, but Hamersley's leadership was doubtful about this, or at least in dealing with its local variant, the Iron Ore Industry Consultative Council. While not as openly opposed as was Robe, the Hamersley management 'paid little more than lip service to its objectives'.[29] Mackinnon's account of the changes in CRA makes clear that 'cost effectiveness and reliability of supply' drove company strategy and that at Hamersley the focus of management action was very much internal, looking at its own operations, not to the wider politics of the industry.[30]

Mining companies were not only fixed in place, as we have seen, but were price takers as the demand for iron ore rose and fell. For many years, these geographic and financial constraints had shaped a company strategy of union accommodation but now, as national politics and local circumstances changed, Hamersley was prepared to rethink its orientation to the Pilbara workforce. As it turned out, the workforce itself brought matters to a head after this period of relative industrial calm.

In June 1992, the workforce struck work over the employment of a non-unionist. This was only six months after the unions at Tiwai Point had collapsed, with almost all the workers signing non-union contracts. The Pilbara unions handed Hamersley's managers an opportunity they were only too willing to grasp. The revelation that a non-union member was employed at Hamersley came about due to an AMWU levy for a national fund to support a tough dispute in the paper mills in Burnie in Tasmania. The Burnie dispute was one of the first in which Robe's strategy was tried outside the Pilbara, and key Peko managers played a vital role in driving the conflict at one of Tasmania's most important industrial sites.[31] The ramifications came right back, disastrously for unions, to the Pilbara. When the levy was imposed, it became clear that Philip Beales, a fitter working at Tom Price, was not a member of the union. Nor would he join, he informed the AMWU convenor. If that was unusual, so too was the company's refusal to direct him to do so or, failing that, to dismiss him. State law had changed since the early days of 'preference for unionists' and, as we saw earlier, the Commission had tried to wind this back too, but nobody had acted to break this understanding of custom and practice at Hamersley. Until now.

What is clear is just how normalised these processes had become in the minds of unionists. For them it was the company's stubbornness over one equally stubborn worker which was putting exports at risk, not the unions' attitude. 'We don't want to stop a multi-million dollar business for the sake of one (expletive) idiot', the union reportedly told Hamersley.[32] The managers, however, were resolute.

When the strike began, in defiance of an order from the Commission to return to work, Beales was joined by two others in staying at work, from 20 June. A handful of other workers also joined Beales in crossing the picket lines at Tom Price. Six days later, the company increased the pressure in what was already new territory with a hand-delivered letter to AMWU organiser John Mossenton, threatening him with personal damages, claiming the company had lost sales worth $45 million. More than a little startled, the union quickly recommended a return to work for 29 June. This was not going to be the end of the matter, because Beales was on leave until 3 July: what would happen after that?

The uncharted territory of this dispute now got even stranger for the unions. The very day of the return to work, the company went to the Supreme Court seeking damages of nearly $50 million with more to come, believed to be the largest such action ever taken against unions in Australia. What is more, the company sought an injunction not against industrial action – after all there was none – but against threatened industrial action. The day before Beales was due back at work, the Court granted this. The union was told it could not force Hamersley to dismiss Beales, nor should anyone interfere with Beales' return to work.

Perhaps unwisely, the union had asked if there was any precedent for such an injunction. And, yes, there was: it was from 1976 and involved a company connected to Jeff Kennett who, now as Premier of Victoria, had just overseen the introduction of the strongly anti-union laws that other states, most notably Western Australia, were about to emulate.

With the unions boxed in and weakened, Hamersley now seized the opportunity to drive home its advantage. Little wonder that some unionists were beginning to think they had been set up by the company, because Hamersley quickly rolled out a series of initiatives that were straight from 'all staff' theory and recent practice in New Zealand.

A harder edge to the company's moves was revealed that had been explicit in the philosophies of Jaques and the thinking of Carnegie. Tellingly, as with Copeman and Robe, the H. R. Nicholls Society was the site where the company story was celebrated through papers by the lawyer Russell Allen, a leading company advocate who appeared in many other cases against unions in the years ahead. Indeed, as 'all staff' practice spread, law firms and key players like Allen became all the more important to companies such as Rio, for whom HR professionals, not industrial relations specialists, were now all important.

Allen made plain just how comprehensive and integrated the strategy was as Hamersley built on its internal rethink and external challenges – and successes. Over the next few months, as Allen later explained, the company aimed to protect those who had scabbed on the strikers. This support was less than those men had hoped for, but it sent a clear signal to the workforce and the community more broadly, as did the clever use of a new word for such workers: they were not strike-breakers but 'independents'. Hamersley dismissed workers it believed were harassing those men. As had happened at Robe, the managers had very little contact with unions, nothing beyond the minimum required. Hamersley had more success than BHP in undoing traditional work practices: the Commission threw out an application from unions to ban staff from doing the work of wage-hands. Arguably, such a win was essential for the 'all staff' plan. The company also sidestepped the unions by providing a wage rise in a way that looked unilateral. Then came the more difficult tasks: not only reducing the size of the workforce, but weeding out high-profile unionists. In Allen's words, 'many of those with the wrong attitude left'.[33] This change to the nature of the workforce anticipated the phrasing in a 'Culture Change' document in 1998: 'Change the people or change the people'.[34]

The redundancies announced on 7 December 1992 were significant in raw terms – at 13 per cent of the workforce – and still more so in terms of effect. Not only did many leading unionists disappear, but so too, as Mackinnon explains, did a cohort of less well-known but older workers, for whom the redundancy packages were particularly attractive. Because these old hands were well versed in local history and practice, this was another change to local culture and a loss of union strength.[35] Put together, as some lamented and others celebrated, all these changes meant that the unions quickly lost their way. Workers lost confidence that unions could actually deliver anything for them. The much-vaunted convenors were, in many cases, exposed as paper tigers: when things got tough they had little to say or offer.[36] According to Allen, the union leaders could no longer even 'convince many employees to attend union meetings'.[37]

As with Robe River, the only remaining barriers to full de-unionisation were the state-sanctioned practices of arbitration, awards and union recognition. Just as Robe had, Hamersley was to take its opportunity to complete the rout. Legislative change was promised by the likely new state government. After the tough stuff – and with those writs still hanging in the air – came the teamwork side of the new deal. Hamersley drove the message of shared values in a much more subtle and comprehensive way than Robe had done. It overhauled human resource management systems, developed new grievance processes, and introduced team-based work from the second half of 1992 and through into the following year.[38]

Throughout 1993, after the state election and pending the new legislation, the company's managers worked hard to offer, explain and sell the new individual contracts which they hoped would bury the old union-based award system. The last date to sign up was 17 December. Things were slow at first but despite union opposition and 'diehards' arguing that loss of collective rights was too high a risk to take, there was 'a sudden flow of acceptances on the last day', with about 90 per cent accepting the offer. There was not much left on the union side after the company's patient, tough and long campaign. Union weakness played out against what the company was offering in its WPAs: salary increases, improved superannuation and health insurance.[39] Subsequent accounts reveal, as later chapters will show, that many workers felt a sense of trust and hope about these changes, ushered in with material benefits and, it seemed, a new kind of management.

The dispute did not entirely wither away, as, like Robe River, it faded from the national gaze. Importantly, the threat of huge costs hung over unions and their leaders. In a very odd twist, some of the 'independents' were in dispute with the company by the early 2000s, talking to union organisers who were trying to 'get back into Rio'. These men claimed that they had had suffered stress, family and personal problems; most quit and drifted away. Two of them were still holding out in legal actions against the company, arguing that in the end Hamersley had not met its duty of care to them. When they tried to take a collective 'fair treatment' case against the company, they ran into a wall. Their managers said that these fair treatments could only be taken on an individual basis. But one, Mark Troy, still wanted his day in court. Another, Phil Randal, now on the east coast, wanted to tell his story. 'Being called a scab', he says, 'it never sat right with me'.[40] As for Beales himself, no trace was found for many years and no answer to the question many people wanted to ask: was he part of an elaborate trap set for the unions in June 1992?

Meanwhile, among the workers who had stayed true to their unions, a small body calling itself WAIC, or Workers Against Individual Contracts, continued to push back against the new way. These workers refused the inducements offered to them and stayed in their unions and on their award, but with age and time their numbers fell. By the early 2000s, they were dubbed the 'dirty dozen', a title they wore as a badge of honour. They supplied union officials in Perth with updates on developments on the ground in the Hamersley sites, and continued to push the company to preserve and improve their award-based working conditions as the industry changed around them. In surely the oddest wash-up to this dispute, one of the staunchest of these workers, Steve Ward, a fitter at the Tom Price mine, was often asked by new workers and bosses for details of the background to problems on the job, and terms in the old award. Some of these workers would still be there, as he was, when the unions came back recruiting in 2002.[41]

To this day, debate continues about whether the state unions, local convenors, or the ACTU could or should have known more, and done better in resisting change at Hamersley. In the face of this new management plan and with one hostile government – the state – the unions' uncertainty was disastrous. Those unionists who argued at the time that this could be a vital dispute over the future of unions, that it was about much more than the fate of Philip Beales, were to be proved right.

New ways of working

The immediate (and lasting) change that followed from de-unionisation was in working hours. Before this, the companies had, somewhat nervously, looked at having two twelve-hour shifts instead of the norm of three shifts of eight hours. The lure of reduced costs, potentially massive savings, especially with 'hot-seat changeovers', was telling. In the loco sections at Robe and Goldsworthy, especially when coupled with changing from two drivers to one, this was, to say the least, a complex set of changes to pursue. Managers had now made life a lot easier for themselves by getting rid of the union agreements which had compelled them to win approval from the workforce for variations to working time.

In this change, the Pilbara was not unique: changes to working time were among the most marked workplace impact of changes to government policies in the 1990s. Even in coal (although not until the end of the 1990s), the militant CFMEU conceded ground on shift arrangements.[42] In the Pilbara, though, the changes were marked, after two of the three companies became 'union-free'. Not only did the twelve-hour shift become the norm, so did a longer working week. During the 1990s, the proportion of miners working a standard week of 35 to 40 hours nearly halved. In metal mining, about two-thirds were working over 49 hours by 1996 compared with just 4.2 per cent in coal.[43] Detailed studies showed that, across sectors, these very long hours were 'strongly related' to the use of individual contracts.[44]

In the Pilbara itself, these changes in working arrangements were not only related closely to changes in legislation and unionisation but to the rise of FIFO. It might be an exaggeration to say that FIFO required the working of twelve-hour shifts, but there was certainly an overlap in terms of the logic of maximising the company's use of labour, especially when flying the workers into the mines.

Having secured more control over the Pilbara, Rio drove change elsewhere. In setting out its vision for employment relations to a government review years later, in 2012, the company argued that leaders 'do not absolve their responsibility and accountability...to third parties', be they other departments in the company or parties 'external' to the company. That, of course, meant unions.[45] Rio's code of conduct, set out in the company's booklet, *The Way We Work*, recognises 'the right of all employees to choose to belong or not belong to a union and to seek to bargain collectively'.[46] Yet, from the time of the disputes at Tiwai Point and Hamersley on, the company spent a lot of time and money developing management strategies, arguing for political change, and running court cases that were *precisely* about privileging one kind of agreement, typically various forms of individual contract, over another, typically union collective agreements.

The whole point of different forms of contracts and workplace agreements is that they do confer different rights and responsibilities on the parties covered by them. The question of guaranteed union representation in grievance procedures is only the most obvious of these differences. In addition, the way agreements relate to each other affects workers' conditions, because some forms of non-union agreement might 'override' the conditions in other agreements. Changing the hours of work was a clear example.

As CRA became Rio Tinto, the influence that this transnational corporation brought to bear across Australia became ever greater. The company had always been watched by owners and managers elsewhere, and was increasingly imitated from the 1990s. The term 'all staff' became 'direct engagement' and was spread by former managers and board members into many other industries, most notably into big companies in banking, telecommunications and airlines, and into think-tanks and lobby groups which shaped federal labour law from 1996. The CFMEU, which, by virtue of its place in the coal industry, was the union most in conflict with Rio, claimed in 2011 that Rio was still the 'flag-bearer for many other companies'.[47] The union's leaders argued that few other companies have had such a sophisticated and comprehensive set of strategies to rework how labour is managed. Rio continued to shape working life in the Pilbara and across Australia for a generation after the Beales dispute.

The legacy of de-unionisation

After taking very different paths, two of the three companies mining the Pilbara arrived at the same juncture at the same time, with political and legislative change allowing them to remake their sites as non-union places. After Robe River's war of attrition and Hamersley Iron's direct assault, both companies 'offered' their workforces individual contracts. In both cases, a massive majority of the workers abandoned their unions. For the employers, Russell Allen was right to celebrate media coverage which argued that: 'this dispute may be geographically remote, but the issues at stake reverberate across the continent'.[48]

At Hamersley Iron as at Robe River, the individualisation of the employment relationship was intertwined with changes in social relationships, changing not only unionism but family and community life. In the immediate aftermath of the Beales strike and the contract offer, the mining towns were divided and unhappy places. Once that period had passed and the workplaces become non-union, there were other and more long-lasting impacts on the towns. As the Pilbara became less and less a union space and less and less a community space, the employers were able to assert more control over the workforce and the workers' time. When coupled with the spread of twelve-hour shifts in the second half of the 1990s and a significant increase in total hours worked, there were marked impacts on family and local life. As all-staff employment morphed into 'direct engagement', the need for workers to identify primarily with the company they worked for became all important. This was obviously the case in over-riding the relationship between worker and union, but also in changing the link between workers and community and even between workers and their families.

The changes that had taken place after the Beales dispute at Hamersley remade much of the Pilbara and challenged the national contours and politics of work and employment relations. Yet the Pilbara had not been wholly re-invented. Australia's biggest mining company, BHP, still faced a seemingly strong union movement determined to maintain its say in the communities and workplaces of Hedland and Newman.

Chapter Seven

'The Last 500': BHP Billiton

BY THE MID-1990S, BHP (on the eve of becoming a still-bigger global company, BHP Billiton) was the only unionised operator in the Pilbara. Since the Robe River de-unionisation had started in 1986, union membership had declined across the country, while from 1996 a federal government had followed the example of most of the state governments in passing laws, notably that year's *Workplace Relations Act*, designed to introduce individual contracts and reduce union power.

Under the Western Australian laws that both Robe and Hamersley had used, managers at BHP had the opportunity to change how they ran their Pilbara sites. They also had a motive: costs were rising and ore prices were uncertain. They were beginning to feel that their non-union rivals in the Pilbara, especially Hamersley Iron, were performing much better than they.

In November 1999, BHP announced it would offer individual contracts to its iron ore workforce: like the other companies, BHP now wished to avoid dealing with unions. Over the next few years, BHP became the site of a sustained campaign by the unions, backed by the ACTU, to hold onto their last ground in the Pilbara. A little over half the workforce quickly walked away from their unions, but a solid core, 'the last 500', stayed loyal. This chapter tells that story, through to 2001 when the union won a new collective agreement for its remaining members. As we shall see in subsequent chapters, the battle would go on, becoming ever harder for the unions and their supporters as the industry changed around them.

Managers and unions at BHP

In the late 1990s, BHP was transforming itself. The company came under the control of a series of high-profile CEOs recruited from outside Australia. It began to shed many of its so-called 'downstream' operations to become a less diversified company, and reworked its once insular and bureaucratic management structures. About 20,000 jobs were lost from 1999, and the share price began a long climb. Among casualties of these changes were the steelworks at Whyalla in South Australia and Newcastle and Port Kembla in New South Wales, which had long and close connections with the Pilbara. There had been massive job losses in those places in the 1980s and, between 1999 and 2001, the steel divisions were sold off. These changes saw BHP concentrating more on resources such as iron ore. The company often known as the 'Big Australian' was in the process of going global. In 2001, BHP merged with the South African company Billiton to form a new company, BHP Billiton.[1] Changes were afoot in other companies too. Rio Tinto was taking over Robe River, with that company joining Hamersley Iron in the Rio stable. For local workers and what remained of their unions, this meant that their employers would soon be the biggest (BHP Billiton) and third biggest (Rio Tinto) mining companies on earth.

These global and national changes affected the companies, life in the Pilbara and the politics of the west. There were also changes in national unionism which had a local impact. Union membership was continuing to fall across Australia, with governments and many employers dismantling the framework that had allowed high levels of unionism and collective bargaining to flourish. Union density (the proportion of the workforce belonging to unions) had been in decline since the late 1970s, but from 1992 the raw number of unionists began to fall. With a rapidly increasing workforce, this meant that density went into freefall for the first time in the twentieth century.[2]

National union strategies designed to address this decline and to deal with growing hostility from employers had been discussed throughout the union movement since the 1980s. As we saw during the Robe River dispute, one proposed solution was a new kind of politics – the Accord between the ACTU and federal Labor governments. Mining employers and many Pilbara workers, however, had been sceptical about this strategy. Another answer, pushed by the ACTU, was union amalgamation. Seeking power through rationalising their resources, federal unions merged and cut their numbers from 146 in 1988 to just forty-five ten years later. Neither strategy had stemmed the bleeding in membership numbers.[3] In the west, the decline was especially marked because of the changes made to labour law in 1993.[4]

Plans conceived nationally, be it in the corporate or political world, or, in this case, union headquarters, often work out differently on the ground, their impacts varying from one place to another. This was emphatically the case in the Pilbara, with the Federated Engine Drivers' and Firemen's Union bringing, in effect, the old Miners' Federation into the Pilbara for the very first time. The 'Federation' was widely held to be one of the country's best organised in workplaces. Until the 1990s, it had often been a communist-led union and remained militant in its industrial strategy. It had a relatively successful history of opposition to Rio Tinto. It can be fairly said that if the mining companies were wary of this move then so too would have been some of the unions, not least the AWU, a union with a very different history and ideology.[5]

In 1999, by coincidence the same year that BHP's managers were gearing up to attack the unions, the union movement adopted a new plan for revival, an 'organising strategy', based largely on developments among embattled unions in the USA.[6] An ACTU booklet entitled *unions@work* explained that unions could no longer simply 'service' the needs of members, they had to enable workers to organise themselves. Successful unions, said the ACTU, were ones making internal changes to use resources efficiently and empower members while, crucially, linking growth strategies to everything else they did.

To return to BHP: in the 1980s there were nine unions in Mt Newman Mining's operations; in the 1990s, despite the amalgamation process, there were still five: the AWU, ETU, AMWU, the TWU and what was now the Mining and Energy Division of the CFMEU. Between them, the five unions at BHP represented practically all the wages workforce. Early in 1999, there were 2,200 employees, of whom about 1,500 were workers with wages and conditions set by an award and an enterprise agreement, referred to by all as 'EBA 3'. There was also, as there had been since the 1970s, an agreement which set out the rights of the unions in relation to the company. EBA 3 was coming to its expiry date, but it started to seem to the unions that the company was stalling on discussions for what would be EBA 4. They were right. The managers had looked across the Pilbara, looked within their own organisation and reached a decision. They had had enough of dealing with unions.

Project Phoenix: Attacking the unions

Since October 1998, BHP had been organising its iron ore operations under the 'Vision 2005' strategy, committed to cost-cutting, reduced employment numbers, and 'cultural change'. In February 1999, the new president of the company's Steelmaking and Energy Materials Division, Robert Kirkby, who knew the Pilbara well, having been a general manager at Newman, decided that they needed to push these changes through iron ore much more quickly. After negotiations in Japan, iron ore prices were cut by 10 per cent and BHP's output seemed likely to drop.[7] As had so often happened, changes in product markets led to internal tensions, but in this case many other factors came into play. BHP's first reaction this time around was an offer of voluntary redundancies to the workforce.

BHP's senior managers began to work on an ambitious plan which they called 'Project Russia', a merger with Hamersley Iron. BHP had long seen Hamersley as their major local competitor because its ores were at the higher end of the market, like BHP's, and sold into similar markets. Through a due diligence exercise, the BHP managers became convinced that Hamersley Iron really was a more-efficient, lower-cost producer. The more they saw of Hamersley Iron's operations, including its 'all-staff arrangements', the more they became convinced that they needed to address flaws in their own operation. That meant a much more effective strategy would be needed than in 1988. At times, BHP's growing concern about the unions' role was quite explicit. When the dispute with the unions ended up in court, Jeff Stockden, the iron ore division's Vice-President HR, said: 'if you could exclude third parties, ie the unions, there was the prospect of getting better flexibilities and therefore greater productivity'.[8] As we shall see, though, anti-unionism was usually encoded in other forms of expression.

The merger idea lasted only a few months. Project Russia morphed into Project Phoenix with Kirkby telling his managers that the iron ore division needed to match Hamersley Iron's performance – and in just 100 days.[9] This was as demanding a timeframe as in 1988, but it went further. The idea of replacing the EBA with individual agreements was on the table as an option from the beginning.

The non-union option looked even better to the BHP management team when union division fractured the single-bargaining unit at BHP. There are many versions of this story. Some say it was simply another instance of the long history of disputes between the AWU and the CFMEU. For others, it seemed that the company wanted to cut out the CFMEU as the more militant newcomer to the Pilbara and deal only with the more familiar AWU. Still others believed that even the AWU had outworn its welcome as far as senior managers were concerned. At this same time, as had happened before in the Pilbara's history, senior managers were looking outside the Pilbara to solve problems within it, in this case meeting ACTU officials, telling them, in a familiar complaint, that the local unions and workers needed to be brought under control. The questions were by whom and how.

The local inter-union tensions surfaced in October 1998 and ran well into 1999 after Don Bartlem, a dynamic, even charismatic, local AWU organiser (who had first come to the Pilbara at the time of the Robe River dispute), quit that union and joined the CFMEU. Would the AWU members come with him into the CFMEU? What was the CFMEU up to behind the scenes? What would all this mean for the orientation of the managers to the unions? Answering the threshold question about coverage rights would take the competing unions back into the state Industrial Relations Commission.[10]

Meantime, BHP's bosses considered their options under Project Phoenix. They were prepared to accept a workforce split half and half between union and non-union workers, between individual agreements and the award.[11] BHP had now decided to take advantage of the state government's regime of individual contracts, just as Hamersley Iron had quickly done, and as Robe River had done after their war of attrition. By design or coincidence BHP's managers followed the CRA/Rio strategy first unveiled in New Zealand and then closer to home at Hamersley. They stalled the negotiations for a new agreement while talking up the need for workplace change and reducing employment levels. They encouraged high-profile and experienced union figures to leave the Pilbara. At the same time, and again just as at Hamersley in the early 1990s, there was a lot of talk around the worksites and in company publicity of 'the new BHP'. Consultants came and went, team meetings – 'hug-and-tug sessions' to the sceptics – were held, and the company's websites were reworked.[12] The Pilbara would be the site of a fundamental struggle over unionism once again.

The contract offer, 1999

On one of Australian history's most storied dates, 11 November (Ned Kelly's hanging, World War I's armistice, the sacking of the Whitlam government), BHP's iron ore workers received their contract offers. If workers signed up within three weeks, they would receive a bonus payment. Someone in BHP's management team seemed to have a good sense of history (or humour) because that acceptance date, 3 December, was also symbolic. It marked the end of a mining rebellion in 1854, the fall of the Eureka stockade.

In this last bastion of Pilbara unionism, most workers' attention was on the present as they and, in many cases, their families considered the contract offers. The offers had their generous side, as they had at Hamersley Iron in 1993, but, of course, once more, represented a challenge to everything the workers had experienced in BHP's Pilbara's worksites and towns. Working life, community and friendships were all on the line. The new contracts were WPAs, the same type of agreement as now in place at Hamersley Iron and Robe River. In the BHP case, the agreements came with cashing-out of sick leave and apparent wage increases of up to $20,000 if workers would walk away from being covered by the unions. They would allow the management to achieve its stated aim, 'the removal of the needs to negotiate change with union representatives',[13] because, as with other such contractual arrangements, any discussions would be between the individual worker and the company's supervisors and HR staff. These agreements contained no provisions for union representation or defence.

As had been the case since 1993, the state law required these agreements to contain only six clauses (compared with much more detail in the award and union agreements at BHP).[14] This was an example of a general trend across the country: shifting rules and procedures from awards and collective agreements to internal company policy. The six matters were: the parties to the agreement; the term of the agreement; the terms and conditions of employment; dispute-resolution procedures; a note that unfair dismissal claims could be referred to the state Commission; and a statement that no threats had been made. Under 'terms and conditions of employment', it simply said that the terms were: 'set out in the Employee's contract of employment and the Staff Handbook as amended from time to time'. The 'Staff Contract of Employment' was, therefore, the document that really mattered. The agreement gave management very specific types of flexibility, by control over time, cost and space. The staff contract delivered hours flexibility through a clause that stated 'circumstances may require you to work outside your normal hours to ensure that the full requirements of your role are met'. Management also 'reserved the right' to move employees from night to day work or from one shift to another. Labour costs could be addressed because salaries were to be reviewed annually and 'adjusted at the company's discretion'.

As to the Issue Resolution Process, if a grievance was unresolved after discussions with the relevant supervisor, then the issue moved up the hierarchy, through a superintendent, followed by a manager and then a vice-president. Finally, 'the employee may discuss the matter with the President of the Company'. Only then could the problem be referred to an external body, the Commission. For unionists, this appeared to be designed to do bring complaints to a halt and to remove any genuine capacity for collective representation.

In responding to the offer, there was, then, a lot for workers to weigh up. The changes locked unions out and shifted the managing of employment relations – the frontier of control – back to management. The BHP workers were in a very different environment from Robe River, where there had been seven years of explicit hostility to the unions, or Hamersley Iron, where a divisive strike had taken place and where the threat of heavy fines had hung over the workforce. So, it was unclear how this attempt at reducing union power would play out.

The offers looked attractive enough in themselves, as they had to be, to win workers away from the collective agreement and, in effect, their unions. The company could afford to make the offer generous, confident that workplace change and efficiency gains would follow. The managers hoped, as had so many others across the Pilbara and around the world, that 'all staff' arrangements would lead to a change of culture – that workers would be more committed to their work and to the organisation. So the plan set out to reward commitment, not with a one-off hand-out but with promises of individual reward for individual performance. If the big stick was to be used to beat off the unions, the carrot was, it seemed, trust in individual ability and trust in the company. Many workers were happy to be part of this, 'the new BHP'. Some workers, indifferent to or disenchanted with their union, were persuaded that a new and better period was signalled by this shift to individual arrangements The union-oriented workers were worried as much by what the agreements did not say, as what they did. For these workers, the minimalist nature of the agreements worried them: too much to risk. Playing on the acronym, WPAs, they dubbed the agreements, and those who signed them, as 'woppas'.

For those loyal to the unions, there was one point of agreement with management: money was not the key issue. For many of these men and women, it was just a matter of loyalty. Like the 'dirty dozen' at Hamersley, their view was uncomplicated: joining a union was a natural part of working life. Most of these workers said they would be powerless without a union. Others felt a debt to history and responsibility for the future, one saying: 'we don't want to leave less for our kids than we got'; another: 'we can't give it all away'.[15]

In this case, the planners had timed their move well. With Christmas coming up, many people would be away, hardly the best time to reinvigorate the unions. By the end of January 2000, nearly 46 per cent of the 1,100 once under the EBA had signed up.[16] Resignations from the unions were rolling in.

Union in-fighting had not helped. In fact, they had been in the state Commission arguing over the dispute between the AWU and the CFMEU at the very time the company was finalising its contract offer. Nor were the unions as strong on the ground as they had been in the 1980s. The workforce was still highly unionised, but, at about 73 per cent, it was not the total union shop of the past.[17] Active commitment to unionism was lower than it had been; communication between members and officials was, on most accounts, poor. The union diehards felt that there was nothing they could do about those they called 'the greedy bastards' who just wanted the money.[18] The real issue was to fix their own backyard, their unions. What could be done?

The unions regroup

The ACTU's organising strategy had been endorsed by the great majority of unions and formed the core of resistance at BHP. In the eyes of many, BHP would become a test case for the national strategy. This does not mean that on 11 November 1999 the unions had a grand plan. Far from it. Even if they had had one, their leaders could not have been confident of turning the tide, after the defeats of the previous decade in the Pilbara and in the light of changes to federal labour law since then. In 1996, the Howard government's new *Workplace Relations Act* made it more difficult for unions to access worksites and force companies to bargain with them. The unions' campaign unfolded in the workplace, towns and courtrooms.

The legal strategy saw the unions argue in the Federal Court that BHP had breached the *Workplace Relations Act* by undermining something that Act purported to provide: freedom of association, or the right to belong to a union. The unions' case was this: the company was 'injuring' workers in their employment because they were members of a union and was offering 'inducements' to resign from a union. On 31 January 2000, the Court delivered an interim decision in the unions' favour. The company could not offer any individual contracts to workers who had not yet signed until the case was heard in full.[19]

When the case was finally heard, the cross-examination of the managers revealed just how concerned they were about Hamersley's efficiencies and union infighting. For the Court, the central question was BHP's response: was it engaging in an unlawful anti-union campaign? The company's witnesses were careful to avoid saying anything about removing unions from BHP's operations. Workers, they said, were perfectly free to join a union. As Rio Tinto had insisted after the Beales dispute, the managers were indifferent to unionism as such: they simply preferred to have their workers on individual contracts. So important was this line of reasoning that it is worth reproducing one of the vital courtroom exchanges.

Union Counsel: What I am suggesting to you is that you know and your industrial relations people know that if you want to enhance your chances or maximise your chances of keeping away from having to deal with unions, the best way to do that is to create a situation where people leave the union?

Kirkby: Never our objective…All I can say is I don't know what individuals are going to do with their union membership, and I was interested in efficient structures for doing business. I can't take it any further than that.

Counsel: You know that they [the unions] can't offer any assistance in relation to any industrial issue, once these people have signed. That is correct, isn't it?

Kirkby: They have their rights as workers.

Counsel: Yes, but they can't call the union in, you can't get an organiser to come down and speak to the manager and say, 'You've mistreated so and so,' or, 'You've mistreated that group of people', or, 'We don't agree to that change over there', you can't do any of that?

Kirkby: That's right, we were looking for a different process.

...

Counsel: You understand, as a matter of commonsense, that large numbers of your workforce would say, 'If I can't have the organiser in and if the union can't help me, why should I pay my money'?

Kirkby: That's their decision.

Counsel: I know it is their decision, but you are experienced enough in business to know and to expect that that would be the response of a large number of your employees.

Kirkby: That, I am not sure of. Maybe you call it commonsense, but I am not sure of that.[20]

The Federal Court's decision was handed down on 10 January 2001, clearing the company of unlawful behaviour. The Court ruled that the offer of individual contracts did not mean that the company was removing a worker's right to belong to a union. There was nothing unlawful about management wanting to exclude unions from discussion over workplace change. Critically, the Court ruled that the company had not interfered with these rights.[21]

There were wider and long-term implications to this, as well as more localised ones. There were spill-over effects from the Pilbara which affected employment relations across the country. The unions had hoped to win this case, as had the Maritime Union of Australia in 1998 when a government-backed plan to de-unionise the wharves had come to grief in the courts partly because the employer's plan had seemed too contrived.[22] Instead, the precedents from the BHP case flowed the other way: it became accepted that membership of a union could be separated from the activities of a union. This 'black letter law' flew in the face of a century of labour law and practice. The ACTU's response to the decision summed this up: it was like saying you could belong to a golf club but not use the course. Locally, regardless of the outcome, the case had bought the unions some time – a precious twelve months as it turned out – in which to regroup.

The unions' response on the ground was shaped by the ACTU's organising strategy. Two vital parts of the ACTU's strategy resonated with Pilbara history: local activism and community networks. Some grumbled that there was 'nothing new' in the organising policy and that it was all American anyway, but the fact that there was something familiar about the union plan was useful. Old ways could be revived. New officials, tied to the ACTU, arrived, and with older ex-Pilbara hands led by the ETU's Jimmy Murie, provided leadership on the ground.[23]

The first steps to set up some local structures to campaign against the company plan had been taken soon after 11 November. The ACTU sent a 'lead organiser', Troy Burton, to Newman and Port Hedland for a two-week visit. Burton was a graduate of the ACTU's training school and well known to leading union officials in 'the east', but he was a Western Australian by birth. Over the next twelve months, as the court case crept on, Burton made half a dozen visits to the Pilbara as the unions began to regroup. Burton later reported that the initial signs were not promising. One of the union convenors who met him on arrival said that he had had enough, threw him the keys to an old car and walked away.[24] Other local leaders quickly accepted that they had made serious mistakes with their infighting and that after their success in 1988 they had drifted along, doing too little for their members.

With Burton as the ACTU presence on the ground, local union members were offered training. Rebuilding active and democratic local structures, rather than leaving things to one or two leaders, was a pressing task.[25] Rather than spending more money in the courts on an appeal against the Federal Court's decision, the ACTU and the mining unions decided it was time to put money and energy into the Pilbara itself. They worked out a deal for joint-funding for a full-time organiser, reporting to the ACTU, to drive the BHP campaign from within the Pilbara.

At the end of 2000, the ACTU appointed a full-time organiser, Will Tracey. Tracey had been schooled and begun union work in Western Australia, although he was born in New Zealand, in

Porirua, near Wellington. So, the ACTU's strategy was actually being carried out by officials with Western Australian experience, not 'wise men from the east'. Tracey was to be a dominant figure in the Pilbara for the next four to five years, admired by the union members and loathed by the company in equal measure. Based in Port Hedland but travelling regularly to Newman (and later into the Rio Tinto towns), Tracey began an exhaustive – and exhausting – campaign. Roads were better than when Gil Barr had done this sort of work for the AWU in the late 1960s, but not much. Distance and demands were no-less intimidating.

It is worth pausing to emphasise how demanding these kinds of jobs were in places like the Pilbara. Tracey was single, with no dependants – and it is hard to see how it could be otherwise because this work was as 'un-family-friendly' as could be imagined. Union officials have been primary targets for many in the media, but, in sites like this, the work was all but ceaseless and without extrinsic reward. There was no gap between work and non-work time; the only escape was when out of mobile-phone range between isolated mine sites and the scattered towns. A meal or a drink in a pub invariably ended up in discussions, often 'robust' ones, about unions. For all the talk of a new model of unionism, the workers in each town typically thought that the union was not spending enough time looking out for them. The ACTU organisers had to contend with this, as would the state officials coming up from Perth. All this for a salary less than that of most of the workers these men were trying to organise.[26]

The unions had begun the dispute looking divided and had quickly been in danger of becoming invisible. The company's glossy brochures and advertisements and targeted meetings with workers were impressive. Once set up in Hedland, Tracey got the unions' story out onto the sites and into the towns with a weekly newsletter, *Rock Solid*, soon appearing. From the beginning, *Rock Solid* tied this dispute to the fate of workers' rights across the country, trying to revive the feeling that, for better or worse, Pilbara mineworkers were at the heart of something bigger.[27]

Union activists set up their own individual meetings and created a website. One of the most popular contributions to the union website was a poem called 'The Last 500' by Nancy Missler, a shovel-operator on the Newman mine. The opening verse spoke of both isolation and determination:

> There is a place far to the north
> Where the true believers have come forth.
> Shoulder to shoulder, a united band
> This is where the last 500 stand.

The union's challenge was to unify and build on this 'last 500'. The union campaigners knew that they had to overcome the divisions which had undercut their position. Within a few months, a network (or delegate structure) had been established which meant that phone-trees and regular meetings could keep these workers in touch with each other. Meetings of the revived combined Mining Unions' Association were held fortnightly with about thirty representatives present. The idea of separate unions seemed to be on the way out. Ross Kumeroa, a popular Maori worker who had come to Australia in 1989 from his home town of Whanganui, had assumed the position of chair of the new combined organisation and was in no doubt that they had fixed inter-union rivalry. 'It has gone', he declared in 2001.[28]

The workplace was not the only site of local response. Town life and mine life, work issues and social issues, had, of course, been interwoven for years. The first issue of the new newsletter, *Rock Solid*, dealt with these kinds of issues, prefiguring the national debate about FIFO which would emerge ten years later. It quoted a Rio Tinto executive as saying they would really like an all-FIFO operation, thus cutting costs and killing off the Pilbara's towns. In contrast, *Rock Solid* argued that the 'company must understand that they can't ride roughshod over the working men and women of the Pilbara, their families and the communities in which they live and work'.[29] Unions were keen to use the mining communities as sources of power as they had tried to do from the 1970s.

Two things happened which marked out the importance of the local communities and which spoke of the nature of the Pilbara as a site of conflict in defining mining spaces. The first was the establishment of a very active network of women. The second was a move into local politics.

The immediate spur to the women's network came in January 2000 after televised violence on the picket line at Newman during a short strike over delays in talks on EBA 4. Some of the women on the Port Hedland picket line – wives and partners of the strikers – began to talk about forming a women's support group. Within a week they had gathered eighty women and their partners for a meeting and barbecue in the town and had established their own group, Action in Support of Partners (ASP).[30] ASP established its own website and newsletter, ran speaking tours and sent delegates to Perth. Colleen Palmer, the Founding Convenor of ASP in Port Hedland, wrote that these women 'felt the need to collectively support their husbands, their partners, their brothers, their fathers, and their grandsons in the stand they were taking against BHP'. She went on:

> We were all a long way from home and living in a very remote and isolated environment that wasn't exactly conducive to family living. We all had concerns about our future in the Pilbara and worried about our children's future and the future of working conditions we were to pass down to them.[31]

Palmer's account of the ASP role made clear how the home and the workplace overlapped, whether the families liked it or not:

> Information including videos, newsletters, drafted plans, and contracts from BHP management was becoming ever more intrusive into our homes, generating uncertainties and insecurities. It was important to the women that we were all as

well informed about the union campaign as BHP intended us
to be about theirs. By doing this, it made the women aware of
what their husbands, partners and family members were being
subjected to every time they stepped on site.[32]

A second local development came two weeks out from Port
Hedland's local council elections in May when the unions
decided to run candidates 'to ensure', said their publicity, that
'families, communities and workers are represented on council'.
Two high-profile unionists, Paul 'Curly' Asplin and Arthur Gear
were elected. The ASP's Colleen Palmer just missed out.

These developments showed that 'community' was not solely
about place or structures. Rather, an individual's decisions were
made in social settings of one kind or another. Mineworkers
were friends with workmates, although some of those friendships
would not survive this period. These workers also made decisions
as fathers, sons, husbands; in some case as daughters, wives,
mothers; perhaps as part of the Newman's Maori community;
perhaps as part of Port Hedland's Islamic community.

In April 2001, after the Court's final decision, BHP had made
a further round of offers, but Kumeroa assured ACTU officers
in Melbourne that there would be no further losses: 'We're
solid'. He claimed that, within hours, 80 per cent of the contract
offers were back in the union office, rejected. None of this had
been easy. BHP's management had not been sitting on its hands:
'information sheets' or agreements were handed out on site; BHP
established a WPA website and ran advertisements on all local
commercial television stations. The unions complained that, 'on
the job', the company was running a 'no tolerance policy' towards
their members.[33]

The final aspect of this early part of the struggle between
BHP and the unions was played out in the Commission, where
the unions were seeking a new collective agreement for the
workers who had stuck with them. All the rebuilding and morale
boosting activities would come to nothing if the unions could

not do the basic job of improving wages and conditions. The company's delay in negotiations for an enterprise agreement and generous offers under the contracts had pushed the unions into a corner. They now had to force the company into the Commission to try to get an 'enterprise award' under state law. In November 2001, the Commission handed down the new award, granting an immediate wage rise of 14 per cent with a further 6 per cent rise in twelve months. Unionists greeted it as a triumph, although there were trade-offs around workplace change, one of the issues that mattered most to the company.[34] Securing the first union-based agreement since the WPA offer in 1999 was in itself a boost to union morale, all the more so because it carried so large a wage rise.

Remaking the Pilbara

From 1999, the Pilbara once again became a testing ground for rival ideas and strategies, this time in a struggle between a global company in transformation and a union movement seeking to re-invent itself. Union resistance was more successful than had been the case at Robe or Hamersley, but this regrouping among existing members masked deeper, ongoing changes in the industry. Fundamental changes in working life were taking place across the Pilbara. The most immediate of these was that, for all their success in holding onto members, the unions were struggling to attract newer workers – a problem that was becoming a national one and which the next mining boom would not of itself solve. The company had declared at the beginning that it would settle for having half the workforce on individual contracts – and they had achieved more than that.

Most importantly, the employers' drive to increase control by making individual agreements was now very clearly interwoven with the rise of FIFO. Whatever motives the companies had for expanding its use, FIFO's key local impact – mostly unremarked – was to remake the human geography of the industry. FIFO compounded the problems the unions had had since the 1960s in seeking to organise a workforce that was fragmented over the Pilbara's terrain. FIFO divided the community and family basis of life from the sites of paid work. The towns had long been troublesome for the companies and, as recently as the 1999 dispute, mobilised as sources of union power. If towns could be done away with, so much the better. Not one new town had been built after the 1970s despite the massive growth in the industry. By early in the twenty-first century, more than half the workforce would be FIFO workers and, with longer hours, it was clear that life, work and employment relations in the Pilbara would be transformed.[35]

By 2001, the Pilbara was for the most part under the control of the companies. Rio Tinto's operations at Robe River and Hamersley Iron were non-union, except for a handful of older workers who had stayed on the old awards. Less than half of the BHP workforce now belonged to a union. When unions tried to resist all this and managers tried to preserve their non-union status in the first decade of the new century, they would do so across a terrain wholly changed from the one over which workers had exercised some control in the past.

Part Four:
Company Space

The defining feature of the early years of the twenty-first century was the Pilbara's most spectacular ever boom, driven by the Chinese economy. Indeed, it is not too much to say that the mining boom often seemed to be the defining feature of work and life across the entire nation. Resources exports expanded to new levels as did returns on investments. The red dirt of the Pilbara, the massive ore trains and airports full of FIFO workers were central to the national imagination of this economic story.

New companies broke into the Pilbara for the first time in forty years despite the huge costs and technical challenges in setting up these operations. Amidst this brave new world, familiar battles were waged. After 2007, changes to national labour law appeared to threaten the non-union systems the companies had set up since 1986. The super-profits being made drove policymakers to look at new taxes. The companies fought back on both fronts. Their newly won hold over the Pilbara was not unchallenged but, even as the boom faltered, it appeared to be intact.

Conflicts between companies and unions were not as vital as they had been in the previous periods this book has examined but they certainly continued. Some local union resistance re-emerged to challenge the power of these global giants but by and large the reliance on FIFO labour as mines expanded, changes in the nature of work and the companies' adroit work in the legal *and political systems kept them in control of the Pilbara. The companies were seeking to make the Pilbara merely a site of production in a globalised network. How that played out is the key to this part of the book.*

Chapter Eight

The Rise and Fall of the Pilbara Mineworkers' Union

THE FINER POINTS of political and legislative change may not seem all that exciting, but both were necessary to the remaking of daily working life and overall employment relations in the Pilbara in the 1990s. One more set of such political changes, against a backdrop of continued change in the nature of the mining industry, lay behind a small-but-significant episode, a localised response to national and global changes in mining. At Hamersley Iron, this took the form of the Pilbara Mineworkers' Union between 2002 and 2004, before Rio Tinto surmounted this union challenge to its most profitable Pilbara operation. Up the road in Newman and Hedland, BHP Billiton also kept control of political and union threats to its newly minted regime of individualised employment relations.

The trigger to all of this came in February 2001, when the Labor Party was elected to form government in the state of Western Australia, promising to get rid of individual contracts. This was a major problem for the mining companies. How could they now maintain the non-union systems they had worked so hard to put in place? The *Labour Relations Reform Act* came into effect on 1 August 2002, phasing out WPAs by 2003. The state legislation had a new kind of individual contract, an Employer–Employee Agreement, but employers were unimpressed: they argued it lacked enough 'flexibility'.

The companies did have a way out. The federal *Workplace Relations Act* provided them with a new solution to the old problem of how to control their Pilbara operations. Simply put, they could escape from the state's laws, taking the regulation of employment relations in the Pilbara outside the state. The federal system gave companies two ways to operate without unions: individual contracts known as Australian Workplace Agreements (AWAs) or a strange form of contract all but unique to Australia, non-union collective agreements, Section 170LKs (usually just called LKs). Rio Tinto was the first of the Pilbara's global corporations to react. It decided to offer its workforce an LK agreement as its way to retain control of its labour at Robe River and Hamersley Iron.

Rethinking unionism in Hamersley Iron

The federal law required a ballot of 'eligible employees' to approve these LK agreements. The result of the Rio ballot, announced on 5 April 2002, was stunning. Robe River workers accepted the deal, but at Hamersley Iron 60 per cent of the workforce voted against it. In the words of Michael Bachelard, an experienced journalist with *The Australian* newspaper, it was not just any kind of shock for managers. It was 'head-in-the-hands, staring-at-the-floor kind of shock'.[1] After nearly a decade of direct engagement between workers and managers, how had this come about?

Like many other employers, Rio Tinto had decided that to manage a large workforce a non-union collective agreement was easier to use than hundreds of AWAs. The company's HR managers were entitled to be confident that they would win workers' support for their proposed agreement. Still, the ballot gave the mining unions an opportunity to put their case to the workers. The unions and the ACTU decided to run a campaign, asking workers to vote 'no'. This would allow them time to win workers' support for a collective *union* agreement instead of the LK. Most union officials in the metropolitan offices were not confident, because they had little time to prepare a campaign and previous attempts to re-unionise Robe and Hamersley had come to nothing. They quickly cobbled together a campaign which drew on the (limited) resources of those around Will Tracey in the campaign against BHP's de-unionisation bid.

The geography of this union campaign had to be very different from others, however, because the unions had no rights to access the workplace. In August 2001, the four mining unions with claims on the Rio workforce (AMWU, AWU, CFMEU and ETU) tried to stake out a different kind of presence. They set up a stall at the Nameless Festival, a fair in Tom Price. As *Rock Solid* put it, the unions were back in 'the heart of Rio Tinto territory'.[2] Local Labor Party parliamentarian Jon Ford reported that many of those at the festival had told him that 'it was good to see a union presence in town'.[3] Generally, however, the union strategists were cautious, believing that most workers and locals had become wary of unions. There was mistrust to overcome as well as the company's usually effective messaging.[4]

The company moved quickly, hand-delivering the proposed agreement to workers' homes across the Pilbara. This was the first Tuesday in March 2002, immediately after Western Australia's Labour Day long weekend. Meanwhile, union pamphlets were printed up in haste, making one key point: there was no hurry to decide. The workers had options and could steer clear of LKs. Voting 'no', the union case insisted, was safe. When the unions did call meetings, over 200 turned out in Paraburdoo and Tom Price, and nearly as many on the coast. This response changed the tone of the company's campaign from a positive message emphasising common purpose to a more negative one, based on their reading of the history of the Pilbara's 'bad old days'.[5]

These meetings revealed why the LK proposal just might founder. There was discontent over working conditions and social conditions, especially in the Hamersley Iron sites and towns. Typical concerns were: shifts being altered with little notice; hours of work being unilaterally increased; and real hourly rates of pay falling. Most of all, it was the processes that are part of individual employment systems that had alienated many: the company's grievance procedures (the 'Fair Treatment Procedure') and performance reviews had aroused a lot of discontent. Fair treatments were slow; all too often, 'nothing happens', said some. One worker was of the view that 'it's designed to wear you down'. As for individual contracts, workers pointed to contradictions that many researchers were also commenting on: the contracts were not really individualised at all; each worker's contract was the same as all the others. What was individualised was the process around them, but even here it was not really individualised: one worker was pitted against two or three supervisors. Management's control ran through all this. Managers, said one worker, 'changed shifts more than undies'. Workers voiced the concerns raised by shift workers in other industries and occupations: the effects on the quality of working life and social life; uncertainty about family time; loss of pay associated with change.[6]

Workplace issues were not the only things concerning resident workers. Like people across the local communities, they were worried about their towns, falling populations, few jobs for locals, declining social life and deteriorating town facilities. From schools to sporting teams, from hospitals to shops, the impact of FIFOs and reliance on contractors was clear and depressing.[7]

How did these non-union workers explain all this? Many looked beyond these individual concerns to a changing style of management. The soft version of dealing with complaints was: 'well, you're still here aren't you?'; the hard: 'if you don't like it, you can fuck off'. Either way, it generated discontent. There was also declining confidence in managers, partly because of the high turnover rate of young onsite managers. Disgruntled workers, especially older ones, believed that in the mid-1990s managers had been committed to a new way of working. By the end of the decade, they said, that had changed. For these frustrated men and women, Hamersley Iron, now often known as the 'Red Dot' after its logo, was no longer such a good place to be.[8] As the company became more obviously global, it seemed to change. Rightly or wrongly, many felt that Hamersley Iron represented 'the good old days'. Rio Tinto, they said, cared mostly about 'value for shareholders'.[9]

What did this discontent mean for the future of unions? This was no simple matter. At first sight, the anti-LK voters looked like classic pro-union workers. They were hostile towards management; many of the most vociferous had a lot of work experience and were intelligent and articulate – natural leaders it seemed. But they also feared that the unions just would not be able to get back into Hamersley or, if they did, they would make a mess of the opportunity. Workers said the result was 'not mainly' a vote for unions; it was 'not necessarily' a pro-union vote. Most were still more forceful. Was the result a pro-union vote? 'No'; 'no way'; 'absolutely not', said different workers.[10]

These LK ballots made it very clear that history mattered. Workers who had never been anywhere near the Pilbara during the Robe River or Hamersley Iron disputes had certainly heard about them. Stories about everything from strikes over ice-cream to who-stabbed-whom in the back came up time and again. Arguments broke out over what had happened with the rise and fall of unions before. Company inductions and informal conversations with managers provided one powerful, anti-union version of history. How the Pilbara's past was remembered was vital to shaping its present.[11]

The workers who were being drawn back to unionism had a firm idea of what they wanted. The message could not have been clearer:

> If we had a union it would be a mining union…that's what we need – one union for the whole lot…the people are looking for something new.

145

> We spent years and years in the unions here in the Pilbara
> fighting each other...if the union was ever to come back here,
> it would need to be an industry union, everybody in the
> industry represented by the one union.
>
> One union'd be the go to stop that sort of shit [demarcation
> disputes]; one union is what you want...if you formed into one
> union, you're a power to be reckoned with.

The ballot result signalled not the end but the beginning of a new fight for the Pilbara. Hamersley's managers organised focus groups and continued with their one-to-ones, trying to understand what had gone wrong. By June 2002, an internal company report into the LK ballot had identified specific grievances. The report recommended further reviews of the assessment system, better pay offers, improved communication and better community liaison.[12] By this time, the unions had already made the running on some of these issues. Where the company promised to set up 'Community Steering Groups', the unions were putting community issues at the core of their campaign. Some of the problems the company faced had arisen because of salient operating issues: any attempt to alter company relations with the local communities appeared to contradict Rio's commitment to the use of FIFO labour.[13]

The most pressing concern for Hamersley was to decide how to regulate its workforce. The managers revisited their options under the *Workplace Relations Act* and chose the one they thought would give them the most flexibility and would fit with the company's ethos. In September 2002 they began to offer AWAs to their workforce, with a 6 per cent wage increase, the only wage increase on the table. Workers had to decide what to do by 28 February 2003. New workers would have to accept AWAs or not take the job, as courts had allowed.[14]

Hamersley Iron: The rise and fall of the Pilbara Mineworkers' Union

The LK ballot had given unions the chance to get inside Rio for the first time in a decade, but turning the 'no' vote into a re-unionised set of worksites was, to put it mildly, no certainty. Many people in the union movement, if they were honest about it, were as shocked by the ballot result as were Rio's managers. It took some time for the unions and the ACTU to develop a response, to organise funding and to settle on a plan to be carried out in the Pilbara. The four mining unions with coverage rights at Hamersley Iron agreed to work with the ACTU, funding an ACTU organiser based in Paraburdoo where there was a strong group of potential members.

This approach drew on the lessons of the rearguard action 'up the road' at BHP, with the novel twist that inter-union cooperation would be given a structural form. This innovation 'from the deserts' was a hybrid – a geographically defined workers' organisation which was also, in effect, at first enterprise-based, in Hamersley Iron. The membership form explained that this new organisation, the Pilbara Mineworkers' Union (PMU), was: 'a grassroots organisation of Hamersley Iron workers that want to have a voice in their workplace and their community... independent of, but [working] closely with industrial unions'.[15] Workers would still belong to their usual unions but the PMU would coordinate all their activities. Behind the scenes, the legal status and form of the PMU was a complex matter but, on the face of it, the local workers had what they wanted: a single union in Hamersley Iron.

Union supporters faced the same question as the company: 'what next?' The plan was to seek a state-based award, as opposed to the federal options, for precisely the reasons the companies disliked the Western Australian laws. They required 'good faith bargaining' and gave the Commission greater power to settle disputes than did the *Workplace Relations Act*. State laws might therefore enable the unions to force the company to bargain with a union.[16]

For any of this to happen, though, the unions had to be a force on the ground. The law was never enough on its own to change what happened in day-to-day work and employment relations. The PMU became active from June 2002 when the ACTU's new organiser, Stewart Edward, arrived in the Pilbara. Troy Burton was back on the scene along with Tracey, who had been busy in the Hamersley towns in the run-up to the ballot, explaining how the unions were rebuilding to try to resist the BHP contract offer.[17] As had been the case at BHP, they needed their messages to be as strong as those the company was putting out. *The Anvil*, under the banner 'forging a better future for [Rio] families and communities', pushed not only on-the-job issues but the community problems so many residents had identified. It also made a vital claim – that despite the common talk of collective action by workers being a thing of the past, it still did happen and it still could work. The 'no' vote and Rio's forced change of strategy were proofs of that.[18]

The first issue of *The Anvil* called on the company to set up Community Consultation Groups. The focus on 'the community' was not just talk. Edward established connections with community groups, town workers and small-business owners in his base, Paraburdoo. He was soon involved in the town's Community Centre – a vital place of local interactions, grievances and knowledge. Townspeople echoed mineworkers' concerns about population decline and services such as education and healthcare.[19]

Through the winter of 2002, PMU members quietly campaigned to win others to their cause and not to sign AWAs. Some pro-union workers wanted a more dramatic gesture to signal the revival of unionism, a feeling which increased as others began to accept the company's AWA offer. Many workers were thinking that with so much uncertainty, they might as well take the money on offer and go back to the non-unionism that many of them were used to.[20]

Under pressure, the PMU's backer, the ACTU, was ready to go public with a PMU 'organising blitz' seeking to talk to every mining household in Paraburdoo and Tom Price. This was timed for the week beginning Monday 11 November 2002, three years to the day since the BHP contract offer. It was a risky option. One constant feature of remote communities or places reliant on one employer is that companies can portray unions as outsiders stirring up problems. The union campaigners made sure that their 'blitz team' was a cross-section of workers: union delegates as well as officials, women and men of different ages.

Watching the blitz at first hand was to think afresh about the geography and history of the Pilbara. Twenty-two delegates from all over Australia were putting into action a strategy developed in the USA. They were meeting in the old union meeting room in Paraburdoo. The room was filled with the past the unions had lost: so abrupt was the unions' departure after the Beales dispute, that the filing cabinets held union minutes and other records untouched since 1994; posters reminded everyone of politics in the early 1990s; there were, tellingly, stickers through which unions had attacked not the global giants but each other.

The blitz went well. Even those who were hesitant about unions were friendly. There were offers of tea, water, a beer (or two), an invitation to cool off in the pool – and one to fix a non-starting bike. The delegates often spent as much time in discussions with partners as with the mineworkers themselves. The overlap of work and family issues, town and mine, came through yet again. Most of those contacted said they would support the PMU.

By the week's end, the PMU had a supporter base in the mining towns. Kerrianne Mills, who used to work for Hamersley (and whose husband Andy still did) was the unofficial treasurer of the PMU in the town. At the final meeting, she thanked the workers she had just met 'from the bottom of our hearts' for their support in overcoming the town's isolation and locking them into a wider cause.[21]

To establish the PMU as an ongoing organisation beyond the campaign, and to respond to the uptake of AWAs, a lot of work was needed. In January 2003, a preliminary committee met which steered the union until its first formal meeting, in February. *The Anvil* kept urging workers to hold off: 'Why sign? There's time'. So there was, but uncertainties remained and the numbers signing up for AWAs continued to grow as the company's deadline of 28 February grew closer, and the promised back-pay grew larger.

Over the long years of union decline, most people had seen individual contracts and unions as necessarily opposed – not least policy-makers and companies themselves, for whom the whole idea of those contracts was to avoid dealing with a union. In this case, though, the PMU accepted that all mineworkers, including those who had signed AWAs, should be eligible for union membership. The argument was that, unlike the BHP case, there was no existing union for the company to break. More than this, many signed AWAs – and the unions endorsed this – on the basis that they could get back onto the promised state award when those AWAs expired. This seemed a reasonable path of least resistance. To cut these workers off would have been self-defeating. By the deadline, 1,152 workers had signed AWAs. Only 139 had not signed, and so, in effect, had switched from the old WPAs to statutory contracts. Nine of the famous 'dirty dozen' remained on the old award, not all of them impressed with the PMU's accommodating strategy.[22]

By April, the union had served its claim for a state award on Hamersley Iron. Despite the low numbers, this was still an achievement, and a novel one: here was a new type of union seeking a collective agreement but embracing a large number of workers employed under individual agreements. They were fighting against a decade of history and a powerful company far from happy with the state's laws. On the job, six months after the blitz, the PMU loyalists were feeling very conspicuous in the workplace. Would the PMU's members stick to the union strategy? Would the company find a way out?

With workers drifting from the PMU and signing AWAs, and the company locked into the revamped state arbitration system, neither side could force the issue. The PMU persisted in its low-key approach, trying to build numbers and force the company to negotiate in the state jurisdiction. For its part, Rio was equally adamant that the state system was the worst possible option.

For the company, there was a way out — one that involved lateral thinking. This was, for Rio Tinto — as one of the most forceful opponents of unions and the award system — to 'go federal' and ask the Australian Industrial Relations Commission to make a federal award for workers who were not under AWAs. The terms of any such award could more or less replicate what was in an AWA. Under the Australian Constitution, this award would over-ride anything the unions might be able to do in the state. The snag was that a union would have to be involved. Even under the *Workplace Relations Act*, all awards and most agreements were made only between unions and employers, not between employers and individual workers. However, there was some reassurance for the company because, in practice, there was no guarantee that union membership would grow just because an award existed.

Early in 2003, the company, therefore, began to talk in secret to the union with widest coverage in the industry, the AWU. A spokesperson later told the *Australian Financial Review* why the company had acted as it had. A federal award would 'mean that the company [would] be able to largely run their own affairs without being hassled by the unions and the WA Industrial Relations Commission. To do that we had to get close to a union'.[23]

For the unions, it was the classic prisoners' dilemma. In the Pilbara, there was a large group of well-paid workers in a state with low and falling union membership overall. The industry's growth prospects seemed unlimited, so a union getting the jump on the others could secure real gains. The AWU still had the right to cover more of the workforce than any other union: it had the most to gain by going alone, and it had a long record of looking after its own members (and potential members) as it saw fit. Despite the apparent unity in setting up the PMU, tensions between the AWU and the CFMEU ran deep, the threat to AWU members on the eve of the BHP dispute still a fresh memory.

The federal award that the AWU and Rio had agreed upon for Hamersley Iron and also Robe River was, said the accompanying memo, a win for all, being the 'first modern industrial instrument...for more than a decade' in Rio's Pilbara operations.[24] The PMU's supporters claimed it was 'no better than an AWA', did not guarantee unions any more voice than under the old regime, and would not do anything in itself to increase union membership.[25] The other unions decided to challenge the validity of the award in the federal Commission, recognising, as had Rio, that if it stood then it would override anything that the PMU could do in the west's system and kill off, in effect, the PMU itself.

When news of the deal emerged in July 2003, the fury of PMU members was uncontained. Not only had fourteen months of work since the LK ballot come to nothing, but the imposition of a particular form of agreement ran counter to everything that federal union leaders and the ACTU had said about workers 'owning' their campaign. PMU members reported that they felt like they had 'targets painted on them' when they went to work, as wary workers and hostile supervisors taunted them with the failure of a new kind of unionism. PMU members claimed they faced constant hostility, ranging from mild abuse to outright victimisation. The jeering drove home for other workers the message that the Pilbara would never be re-unionised.[26] Vic Davis, a key figure among the rail section, lamented that the position he held in the PMU was now 'acting president of stuff all'. More importantly, like other some other older workers, he felt that national union leaderships had done too little to hold onto the Pilbara, as had been the case in 1993.[27]

The PMU brought the dispute from the 'desert' to the city in August 2003 when two iron ore workers, Danny Connors (from

Port Hedland) and Kevin Quill (from Paraburdoo), addressed the ACTU's Congress in Melbourne. They argued that, although progress had been slow, they had built a platform for the future and had been 'betrayed by our own side'. AWU federal secretary Bill Shorten put the position for his union: the task was to do what the PMU, after months of work, had failed to do: 'to get in the gate'.[28] His union, however, actually could make a deal with the company; the other unions and the PMU had nothing. ACTU leaders called for unity to be maintained. When the PMU leadership argued in a meeting with ACTU leaders that they had only been carrying out ACTU policy, they were rebuffed: 'don't give us that pure organising bullshit'.[29] Political reality trumped stated strategies.

The unions backing the PMU did not give up right away. They took the matter to the state Commission and the federal Australian Industrial Relations Commission. In a bitterly contested case, they argued that the federal award should not go ahead, or if it did it should not bind the 10 per cent of the workforce who had not signed AWAs. In June 2004, the Commission made its decision. It noted that the evidence showed that the AWU's federal officers in meetings with the ACTU and other unions had 'failed to disclose' that negotiations were at 'an advanced stage by April 2003'. The Commission ruled there was no lawful obstacle to the new federal award.[30] The PMU was dead at Hamersley. Edward's term as an ACTU organiser came to an end. He left the Pilbara and was not replaced. Tracey was still mostly engaged with the BHP Billiton campaign.

As soon as the decision came down, Gary Wood, the state secretary of the miners' division of the CFMEU, Jim Murie of the electricians, and Tony Lovett from the AMWU set up meetings with Tracey and PMU members in Tom Price, Paraburdoo and Karratha. Attendees waded through the complexity of regulating wages and conditions in the Pilbara, from the old woppas, statutory contracts, AWAs, and state awards at Hamersley to the proposed federal award (so much for the 'deregulated' industrial relations that politicians were seeking). Murie explained to members the good and bad times he and other Pilbara union veterans had seen come and go, that lessons could always be learned. Tracey cut through the legal complexities: 'It's a shit sandwich, guys, but we're going to have to cop it'.[31]

The workers endorsed the plan that the other three mining unions be made parties to the new award, accepting that it was better to be in the tent than outside it.[32] Sadly for unions, no new organising campaign followed; not one worker was recruited to the union cause. The disillusionment was profound, with more than one person involved commenting that, as with the Robe River dispute, Pilbara unionists, including AWU unionists, had been hung out to dry as national rivalries played out locally. Not only had the PMU vision died but, once again, years would pass until any kind of breach would be made in what was once again 'Rio territory'.

The union rearguard at BHP Billiton

Faced with the same political problem as Rio Tinto, the management of BHP Billiton's iron ore operations decided that it would do what Rio had done and use federal labour laws in the Pilbara. After the ballot result at Rio, the LK route seemed to present difficulties. If Hamersley Iron's non-union workforce had voted 'no', then it was unlikely that the split workforce at BHP would give the company a better result. AWAs were, therefore, the best option. Contract offers were posted to the workforce in September 2002, five months after the LK ballot result at Rio.

If the management at BHP Billiton had problems, so did the mining unions. The company's AWA offer did not worry the majority of union members too much because of the award wage rises they had won in 2001, but they felt that the company was toughening up on them. They believed that changes to rosters were designed to isolate key union members from other workers. They looked with anxiety at the Commission where the company succeeded in barring Tracey from the workplace by having his inspection permit revoked for six weeks after a confrontation with a manager.[33]

The biggest problem the unions faced was a much wider issue – the long-term impact of AWAs. The courts had been saying for years that an employer could not force an existing workforce onto AWAs, but they could insist that 'new hires' sign them. This meant that over time a workforce could be put onto individual contracts without the major confrontations which had taken place at Robe and Hamersley. As older – and more likely union – hands left the industry, the unions would inevitably be left with an ever-smaller proportion of the workforce. The greater use of contractors and the establishment of new mines around Newman as pure FIFO sites did not help either. By the end of 2003, union members, 'the last 500', were down to fewer than 400, about 38 per cent of the wages workforce.[34] Unions needed to reinvent themselves.

The unions' response came in many forms, from the local to the global. A new community organisation, drawn from across sections of the mining town of Newman, emerged after a town meeting on the FIFO issue in February 2003. The Pilbara Sustainability Taskforce (PST) was established after that meeting. Two Upper House members of the State Parliament, Jon Ford (Labor) and Robin Chapple (Greens), had attended the initial meeting, but they thought that local people should take the lead. Some of the speakers at that first meeting argued that by implication if not design FIFO reduced union power and community power against the mining companies. Other speakers talked about losses to local business, the lack of local job opportunities, and threats to the quality, and even the survival, of local schools.[35]

A committee was set up, drawn from broad interests in the town. Sharon Thiel, who became its most public figure, chaired it. Her husband, John Johnston, had been one of the key players in the revitalisation of the Newman unions with Tracey and Kumeroa. The PST campaigned around FIFO and other community issues, pressuring the state government and local council over resources and facilities. For the PST, the Pilbara was about home and family as much as work, about a social setting as much as a mining region.[36]

Meantime, union officials and members had to face the threat posed by the 'new starts' on AWAs. Tracey and others set out to win support for the controversial proposition that the local unions should encourage the 'woppas' to join – and in many cases, of course, re-join – their union. They argued that, in this hostile environment, the unions could not afford to be pure and only organise and defend their own. To do so would leave them with an ever-smaller proportion of the workforce.

Most rusted-on union members could see the logic of this argument, but it was not easy to accept. It broke with the collectivist principles that had underpinned the growth of unions across the world and, markedly, in the Pilbara since the 1960s. When the union men and women looked at those who had crossed picket lines in the strikes of the summer of 1999–2000, it was harder still. For most, 'scab', not woppa, was the preferred description for such workers, often old friends. By the end of 2003, though, union convenors at BHP Billiton had agreed to this position, and early in 2004 the members followed.[37] Despite the formal acceptance of this approach, a fundamental question remained: could unions attract new workers, or had the ground shifted from under them irrevocably? Was a new union approach needed?

Just as the initial resistance at BHP from 1999 had inspired some people to believe in union revival at Hamersley, now the PMU example might re-ignite the union campaign back in Hedland and Newman. The response among Pilbara unionists, including AWU members, to the collapse of the PMU at Hamersley Iron took state and federal union leaders by surprise. Most of them, sympathetic or not to the PMU, had not seen beyond the old union factionalism and politics to the real impact on ordinary workers who had put their faith in a new model of unionism. Perhaps sensing their mistake, they were all ready to support (or tolerate) a new body to give form to inter-union cooperation at BHP Billiton. This was a plan to establish the BHP Pilbara Mineworkers' Union (BHP-PMU).[38]

It took some time to negotiate the BHP-PMU plan in state and federal union offices, where there was some unease about this latest manifestation of Pilbara independence. Tracey and local activists argued that setting up the BHP-PMU as a combined union shopfront would give unions a clearer identity as they addressed their biggest challenge, organising new workers on AWAs. For many workers, younger ones in particular, existing union structures often seemed irrelevant or unintelligible. The aim was, in the words of the inter-union agreement, 'to maximise union membership and improve the situation for workers and the community'; the means: a structure inspired by what had been attempted at Hamersley.[39]

The BHP-PMU had to deliver something to its members after failing to do so before the PMU implosion at Hamersley Iron. In 2003, the constituent unions made a number of claims to try to close the gap in conditions between their members and AWA workers. The last tranche of increases under the 2001 award were now falling behind AWA earnings. In July 2004, the Commission awarded an immediate 8 per cent wage rise with a further 4 per cent on 1 September 2004. The second increase was in line with an expected increase in AWA rates. All this was on top of the increase of 20 per cent in the previous award, a rise which had driven up AWA rates too, it seemed.[40]

The unions could argue, therefore, that it paid to be a member; but there was more to the unions' campaign than this, and more to the judgement than this. This decision attracted little discussion outside the Pilbara – perhaps the west was too far away after all. It was an intriguing case about equal treatment, encapsulating a twenty-year debate about awards, individual agreements and efficiency. The unions argued that it was not true to say that all AWA workers were more efficient than all award workers. Critically, the Commission decided that 'the productivity of award employees and their contribution to the performance of BHPB, given the difference in working arrangements, [was] not significantly different from that of AWA employees'.[41]

The Commission accepted that there should be a 'premium' for AWA employees because of what had been conceded by these workers to earn more (basically, less control over their own work). The Commission estimated these 'structural differences' between an award and an AWA to be worth 8.5 per cent. It found that the wage gap between workers was 16.5 per cent and therefore decided that the wage increase should be the difference – hence the 8 per cent.[42] The decision also noted that 'significant differences in remuneration between employees performing similar work may be a cause of friction or conflict in the workplace'. So, while the Commission did not want to stop the company making AWAs attractive to employees, it ruled that if workers wanted to be on awards, bargaining collectively, then they should not be treated less favourably overall. The Commission did want a reduction in 'the differences between employees who perform similar work'.[43]

In the course of the hearing, the grounds for BHP Billiton's enthusiasm for individual contracts became clearer. In 1999, the focus had been on unionism itself and the 'inflexibility' of the award, whereas they now said it was 'the attitudes of some employees' which were the problem. These workers lacked commitment to the company.[44] The hearings also made plain the gains the company made from individual contracts: the power to change shifts, to determine salaries and conditions unilaterally, and the ability to change procedures as they saw fit.[45] These were the very reasons why unionists had opposed the original WPA push at BHP in 1999. In short, this amounted to employer-defined 'flexibility' – the reason many workers and most unions had opposed individual contracts in the west from 1993 and federally from 1996.

Winning improvements in wages and conditions for BHP Billiton's remaining union workforce was an impressive achievement. It was, however, a limited one. Outside the small union tent, the workforce was beginning to grow at both Rio Tinto and BHP-Billiton. On the eve of the Pilbara's biggest-ever boom, the cold facts were that unionism was but a rump amid a growing, well-paid, non-union workforce. Union revival had never got started at Robe River and had collapsed at Hamersley Iron.

The unions were barely hanging on at BHP Billiton. They remained strong in the rail section, despite almost all of the membership being on individual contracts of one sort or another, but the PMU did not reach out successfully to the new workforce. The promise that had been shown with 'the last 500' proved to be a mirage. Many local people who had been leaders in the union revival and the town's political life at Newman moved away for family or other reasons. A lot of community memory went with them. Not long after the PMU debacle at Hamersley, Tracey's term as ACTU organiser came to an end. He was not replaced.[46] For all the good intentions of building a local network, it was clear that the kind of leadership people such as Tracey provided was critical. The ETU's Murie carried a lot of Pilbara history in his head and was well known across the Pilbara but was based in Perth where most of the members of his union worked. The other unions had fewer such connections. Most importantly, there were now no full-time union officials left in the Pilbara's mining areas.

BHP Billiton had weathered the storm and more than achieved the goals it had set itself in 1999. Its managers, like those at Robe after 1986 and Hamersley from 1993, had remade the workplaces and towns in their own image.

Geography and power in the Pilbara

As the employers' power became more concentrated through Rio's control over, and integration of, Robe River and Hamersley Iron, and through BHP Billiton's still-greater global stamp, the unions and their backers tried to find power in ever more localised and, to some extent, community responses.

In the end, the PMU's focus on community issues in Tom Price and Paraburdoo was as much a weakness as a strength, because it generated tensions among the Hamersley workers. This was because of the changes in the geography of work which employers had made over the years. Many FIFO workers (especially among train drivers) were sympathetic to unionism, but wanted more immediate attention to wages and conditions. 'Community' might have mattered, but for them it lay hundreds of kilometres away in and around their homes and families.[47] Similarly, the local focus of the campaign meant that the unions had done little to explore and challenge Rio Tinto's corporate profile and agenda, something that other community and protest groups were doing. This attempt to build a new kind of local union had no answer to national union power plays. For a short time, at BHP Billiton, the unionised workers, the local community and their rebuilt union structures were a source of power. They could not, however, reclaim anything like the power they had once exercised as the workforce changed around them.

For the union men and women who had done so much for the PMU at Hamersley and then BHP Billiton, the contours of the workforce would change still further in the coming boom, threatening what remained of union power still more deeply. The geography of working life and employment relations was to be changed once more, not as had happened before because of a slowdown but, this time, in the biggest mining boom the Pilbara had ever seen.

Chapter Nine

Workers in the Boom

By 2004, the two big companies had contained, if not eradicated, the union revival which had threatened their dominance of the landscape, just as the Pilbara entered the most marked expansion of the iron ore industry since exports began in 1966. The market uncertainty and cost pressures, which at various times in the 1970s and 1980s had driven managers to try to increase their control and reduce union power, seemed to be things of the past.

After the year 2000, iron ore prices began to increase, mostly because of Chinese demand. The price rose from US$25 a tonne in 2005 to about US$190 at the peak in 2011. One industry source pointed out that between 2000 and 2015, even with a slide in price after 2011, the average price of US$74 was 'more than double the average from 1966 to 2000'.[1] This, then, was a period of major change in the global economy to which the Pilbara was bound. One outcome was that for the first time since the very early 1970s, new companies were able to gather up the capital required to mine the area.

China was not the only force at work in shaping the Pilbara. Complicating things for the buoyant companies were federal politics and at least one group of local workers. The old issues of control, work and employment relations were to get back onto the agenda.

The Pilbara and the boom time

The growth of the Chinese economy was probably the fastest recorded in the globe's history. Driven partly by the movement of millions of internal migrants from rural regions to cities and the Chinese government's commitment to raising domestic living standards, urban construction put immense pressure on the country's steel mills. Australian iron ore and, notwithstanding growing environmental concerns, coal were therefore in huge demand.

Iron ore production in Australia, almost all of it from the Pilbara, more than doubled in the six years after 2006 to the height of the national mining boom, reaching 454 million tonnes, only to keep on increasing in the so-called 'downturn' to nearly 719 million tonnes in 2014–15. The Pilbara's iron ore revenue hit $74 billion in 2013–14. Employment numbers rose to almost 60,000.[2] As the industry expanded worldwide, the Pilbara's proportionate stake in it grew: by 2013 about half of the world's seaborne iron ore market was coming out of Australia.[3] Up to 60 per cent of the country's export income now came from mining, yet less than 2 per cent of the paid workforce was in mining.[4]

The Pilbara became a cash cow for the already massive transnational mining companies; profits almost beggared belief, in some cases propping up other struggling divisions of the two global giants. Ships took over 100,000 tonnes of ore at a time out of the ports to China. The number of zeroes involved in tonnages and revenues soon became hard to comprehend. Perhaps the wealth in the boom is clearest put like this: a single train load in one of Rio Tinto's standard 234-car sets, carrying about 23,000 tonnes of ore, was worth over $4 million at the height of the boom. Little wonder no-one felt overpaid.

So strong was Chinese demand that it shielded the Australian economy from the worst effects of a global recession which tore through North America and Europe from 2008. This was not the only factor limiting the recession's impact, because the Labor government elected in 2007 adopted an aggressive spending policy seeking to avoid recession. For the Pilbara story, the relevance of this was that the episode brought into sharp focus the complex relationship between resources extraction and the rest of the economy. Critics argued that the mining lobby routinely exaggerated the downstream implications of the resources sector and surveys certainly showed that most people had an inflated idea of the numbers employed in mining.[5] At the same time in Western Australia, many lamented the cost pressures building up in the capital city, the lack of offshoots from mining and other jobs and the apparent indifference of policy-makers to the question of what would happen post-mining. In other countries, researchers were thinking about the 'resources curse' around all the downsides to society, environment and economies that minerals dependence could bring. For the most part, media and policy-makers carried on in wilful ignorance of all this thinking.[6]

Meanwhile in the Pilbara itself, a new iron ore operation was beginning, led by a flamboyant entrepreneur, Andrew Forrest, a descendant of one of the most significant figures in the state's history, explorer and first premier of the Colony, John Forrest.[7] Like Lang Hancock, Andrew Forrest and his family had pastoral holdings in the Pilbara, so he too bridged the gap between the area's two great export industries. Unlike Hancock, who had once championed sterilisation of 'half castes',[8] he had close personal relationships with Indigenous peoples and would employ more than most – although his firm's bargaining over royalties was the source of much controversy.[9] Because Forrest had had some truly spectacular failures before the Fortescue Metals Group (FMG) set out to mine the Pilbara, there was more than a little scepticism that the company would break the duopoly of the two big firms. Nonetheless, in 2003, FMG shipped its first ores (having had to build its own rail lines after the courts disallowed its attempt to use BHP Billiton's infrastructure). By the height of the boom, four major mines were operating in two 'hubs' to the north of the Hamersley and BHP sites. The company was shipping an extraordinary 160 million tonnes per year, having achieved its goal of becoming 'the new force in iron' in the Pilbara.[10]

As to the work and workers, little was said in media coverage of the FMG venture. In itself this showed how much things had changed since the industry was set up. Back then, social norms as well as labour law and most company practice mandated that union-based collective agreements would be the basis for setting wages and conditions. All that, it seemed, was gone. So too was the community basis for work. FMG was set up with a FIFO workforce, all of them working under statutory individual contracts. Forrest and others spoke of the FMG 'family', the language uncomplicatedly fitting into management thought far removed from traditional Pilbara mindsets. The only public discussion of the natue of the workforce was around race. Despite tough and divisive negotiations with Indigenous landowners, the company moved quickly to try to increase numbers of Indigenous workers in the industry, who made up over 12 per cent of FMG's total workforce by mid-2013.[11] The single, small break the unions made was among the workers who were in the rail section, (interestingly and as we shall see) the main remaining unionised part of the workforce elsewhere in the Pilbara.

The mining boom generated not only billions of dollars in cash flow, but a debate about the distribution of its benefits and the nature of its legacy. The Labor government elected in 2007 picked up on this and planned to tax what it and others referred to as the 'super profits' of the corporations and then set up a sovereign wealth fund. Chief among the models they had in mind was Norway, where such a fund, based on revenue from offshore petroleum, had driven massive infrastructure developments beyond the norm for so small a country.

The mining companies made clear their political influence in 2010 when they successfully attacked this proposal. Rio Tinto, BHP Billiton and another global operator, Xstrata, ran an expensive media campaign against the government, backed by AMMA and the Minerals Council of Australia. They destroyed the original tax proposal and secured an alternative under which they paid nothing new. In the process, they added to existing pressures on the prime minister, Kevin Rudd, such that his colleagues replaced him – and 'his' tax. Other Pilbara operators were unimpressed with any kind of tax. Forrest, and Lang Hancock's daughter Gina Rinehart, who had become the country's wealthiest woman through Pilbara royalties, led noisy rallies against the tax. They and others painted this, as had the 'boosters' and provincial localists of old, as an attack on 'the west', turning a matter of class and economics into political geography.[12]

In a book about work and workers, the important parts of this story are about geography and history. Some of the arguments against the tax misread the geography of the industry, be it wilfully or in ignorance. With the companies mining high-quality ore at low cost and with relatively close proximity to China, there was no way they were disappearing from the Pilbara in protest. The fixity of the ore bodies locked them in, as it always had. The idea that Australia was now an unreliable place in which to invest – suddenly everyone was an expert on 'sovereign risk', a term few people had heard of till then – was curious. One only needed to think for a moment about the violent, corrupt and unstable places where these companies were mining to realise this. That the companies would seek to deal with the limits geography placed on them, and advance their interests by arguing against a tax, was hardly surprising. That so many other people would parrot the same lines was disappointing.[13] It was hard not to think we had sold ourselves – and the Pilbara's legacy – short.

In the meantime, the social and human geography of the Pilbara was not much improved by the boom. Beneath the money and the hype lay a different set of stories about local economies, drugs and violence, the marginalisation of women and, still, of the Indigenous population. Despite a 'Pilbara cities' program, local politicians complained about the lack of investment back into the Pilbara. Neither housing availability nor infrastructure kept pace with the boom. As labour markets were transformed, so was local real estate, with rents for a basic home in Karratha commonly fetching $2,000 a week. The boom put the Pilbara, the FIFO worker and the mind-bending figures around the industry at the centre of national media attention and public conversation. On the fringes, both literally and metaphorically, lay the descendants of the first inhabitants of the Pilbara from whose lands the profits flowed. It was impossible to reconcile mining's investment levels and profits with the conditions of many of these people. On the outskirts of towns generating the highest profits ever recorded in Australian mining lay abject poverty. It was ten minutes' walk, not a ten-hour flight, to the Third World.[14]

The massive, fenced-in, TV-monitored FIFO camps isolated the new workforce from communities, often because the camps were built much nearer mine sites than to the towns. Some even had their own airstrips. When FIFO workers from camps did mingle in the towns, there were often tensions between FIFO workers (mostly men) and local workers and families. Alcohol played its part in fuelling these strains, but so too, for all the testing onsite, did illicit drugs. Testing merely redirected decisions about the drug of choice.

Not a lot of thought was given to the impact of a workforce of highly paid young men, far from home, on local communities – or even of this kind of life on the men themselves.[15] If drug supply was a boom industry, so was the supply of sex: at least one brothel owner reported that he could not get enough women to work the hours needed in Port Hedland. All this, the dark side of the boom, received some academic and media attention but little policy action.[16]

As to the history, which could have shaped an understanding of the boom and the tax debate in particular, an amnesia had developed that suited the companies very well. Nobody seemed to recall that the industry owed its very existence not only to the ancient ores but to government action, concessions and support. The way the debate played out, although no-one remarked upon it, was another testament to how the companies had remade the

Pilbara. Meetings in the mining towns and company messages to workers about the tax proposal were all but unchallenged. Despite most union leaders supporting the tax, many of the few remaining local unionists were opposed. It is simply unthinkable that a vibrant local union movement such as had existed in the 1970s would have allowed this. They would have engaged in a debate about the tax and in all likelihood put a case against the companies' position.[17] The mining transnationals had asserted their power over Pilbara workers and communities as surely as they had over government policy makers. Their influence over each space – Pilbara and parliament – increased their dominance in the other.

The interweaving of national politics, global companies and the iron ore industry had been made plain by this episode, but this was not the only way in which these things were tied together. The federal election in 2007 which had brought Labor to office under Kevin Rudd, the prime minister who was undone by the tax debacle, had been about one major issue: labour law. The voters had demanded change. The government gave it to them in 2009 with the *Fair Work Act*. Among other things, this abolished AWAs and LKs. So, where now for mining companies' labour strategies in the Pilbara?

National politics and local industrial relations

To the surprise of anyone outside the Pilbara who noticed, Rio Tinto agreed in 2010 to bargain with a union. The PMU interlude aside, this had not happened since the Beales dispute nearly twenty years earlier. And that union was Rio's bête noir: the CFMEU. This news seemed to come out of nowhere, but the back-story lay in part in an unintended outcome of the very success the companies had had in shaping national politics.

The threat which the PMU had posed to Rio Tinto's non-union stronghold at Hamersley Iron had arisen in part because of changes to labour law in the state of Western Australia. Part of the company's successful response had been to escape from those laws and join the federal system of industrial relations. This saga, however, quickly took a new turn. In the federal election of that very year, the returned and empowered Howard coalition government introduced changes to labour law which it had wanted to make for many years. At first sight, the changes seemed to suit the mining companies, but the politics soon turned sour for them and the government.

Under the banner of 'Work Choices', the new laws were more far-reaching than many people had expected: protections against unfair dismissal were removed for most workers; contracts did not have to pass a 'No Disadvantage Test' relative to awards; the law limited what could be agreed upon between employers and employees; it was made easier for employers to back out of agreements; it was easier, too, to close down businesses and re-open with the same staff on reduced conditions; union rights to access workplaces or to strike were still further hemmed in. All this and more in 1,700 pages of legislation in the name of 'deregulation'.[18]

The government was stunned by the public backlash. Part of the problem was that the stripping away of minimum standards was most marked not in export sectors which were competing globally, but in the female-dominated, low-paid service and care sectors.[19] Partly for this reason, the long union campaign against these laws succeeded in unseating the government in 2007.[20] Not even a prosperous economy could distract the voters from the unions' 'Your Rights at Work' message. Although union membership was at historically low levels, Work Choices was the vote-changing issue. The only exception to this trend was in the very state where iron ore was mined, Western Australia. Perhaps this was because individual contracts had been around in state law since 1993. The boom mentality of the early 2000s certainly was a factor. Women's work had been affected early on, but now the main use of the contracts was to control work (and workers), not cut costs. That the union campaign did not bite in the west was telling: Western Australia was, as many locals insisted, 'different'. The booming resource economy and the demonising of unions over the years in the Pilbara were a big part of the explanation for that.

The national backlash against Work Choices nevertheless had a local impact because the system of individual contracts so central to the way the Pilbara now worked was under threat once more. The Labor Party promised to abolish AWAs and restore collective bargaining. Once Labor was elected, neither state nor federal systems would allow individual contracts under labour law. (Common law contracts could still be made, but these were more hemmed in by law as to how conditions could be changed.) The mining companies, especially Rio Tinto, and their main lobby group, AMMA, had of course been among the leading proponents of these laws ever since Robe River in 1986. Throughout the Work Choices period, they continued to urge the government on, attacking union, academic and Labor Party criticisms and counter proposals.

As the Pilbara boomed, the end of Work Choices and with it the AWA era posed yet another challenge to the companies, allowing the possibility that the unions might get back in the gate and into iron ore workplaces. Even before the election, though, the mining lobby secured concessions from Labor, the key one being a long phase-in to any new system: existing non-union agreements would continue until at least their nominal expiry date.[21]

The leadership of the mining companies and their employment lawyers could no doubt read an opinion poll as well as anyone else. That Labor was likely to win the election was clear a long way out. Rio came up with two pre-emptive ploys to protect the non-union status of its worksites. First, it made a new non-union collective agreement; second, that agreement was made with only a part of its workforce, defined by the time of hiring. From July 2008, this Pilbara Iron Employee Agreement (PIEA), would operate for a minimum of five years. It applied to all new employees, based on an agreement with just ten employees who had begun work on a particular date. The CFMEU claimed that the agreement would cover 60 per cent of the workforce by 2013 because of the industry's high turnover and fast-growth.[22] They could see the likely effect matching Rio's intention: locking the unions out for the immediate future whatever legislative change Labor might make.

Meanwhile, at BHP Billiton, there was a core of union loyalists who were still members and who remained covered by a union agreement. In only one part of the operation, the rail network, did anything like a majority of the workforce belong to a union, the CFMEU. In Rio Tinto's Pilbara operations, a few workers remained in a union out of conviction but many of the PMU activists had gone to work elsewhere or were living in Perth. Only in the rail section, perhaps inspired by their colleagues at BHP Billiton, was there much interest in unions. On the eve of Labor coming to office, Rio Tinto employed about 330 train drivers and seventy car examiners. The drivers piloted the 250 kilometres of track between mines and port; the examiners 'walking the train' (checking the locos and ore cars for signs of damage) and 'rolling the train' in and out of dumpers.[23] The specific union to which these workers would turn was, as was the case at BHP Billiton, the CFMEU.

Since the 1990s, the CFMEU had been engaged in what the federal Industrial Relations Commission famously called a 'battle of the titans' with Rio Tinto in Australia's eastern states.[24] Now it was, once more, the Pilbara which would be the site for a dispute over union recognition, this time in the heart of the iron ore boom. How this might play out was by no means clear cut because, despite changes in the law, Rio had a long track record of success against Pilbara unions while the union was, in local terms, no 'titan' at all. It had only a handful of workers and one full-time official, state secretary Gary Wood, based in the far south of the state in the union's coalmining heartland.

The emergence of a local unionism – the rail depots

Why would so well-paid a group of workers want to go through the trials of trying to get a union agreement at all, and especially in Rio Tinto, the leading force in direct engagement? Much of the answer sprang from an issue that went back at least as far as the Robe River dispute: control. In this case, the issues about which workers were concerned were the work process itself and the company's power over workers' remuneration. The immediate spur to action about these concerns was the prospect of changes to labour law after the federal election of November 2007. It was certainly the case that over the summer after Labor's victory, the interest in union membership grew. About seventy workers, along with state officials and some members of the public, attended the first meeting of a Lodge (the term coming from the union's coalmining background) in Karratha in February 2008.[25]

The growth of the union was not, though, simply a by-product of the potential legislative change, nor, as people often see it, as a result of outside intervention by union officials. It was based on the ground in Rio Tinto's Pilbara sites. The rail workers picked up on tactics used in workplaces in the recent past to build the union's profile in the Rio Tinto fortress. As had happened in other organising campaigns, they set out to understand their own workplace not just as a rail depot but as a network of would-be unionists. So, they mapped the workforce and identified likely leaders. The appearance of union newsletters in crib rooms and the wearing of union badges on Rio Tinto clothing made the union visible again. Unionism was no longer just an abstraction or a media bogey.[26]

'Getting back into Rio', as many of the workers and union leaders described it, was a well thought-out campaign shaped, from the beginning, by the way in which Rio managers would react to re-unionisation and by the peculiarities of trying to organise the workforce in this particular place with its striking history and geography. Even when new union members were involved, the workplace itself was not always the site of their activities. Organising a workforce which included FIFO employees meant finding new places in which to meet workers – the camps or quarters where they lived while on roster, or the major transport hub, Perth airport. This would not be easy. As we have seen, the rise of FIFO was the decisive change in the nature of life and work in the Pilbara after the first rounds of de-unionisation. The companies' control of access to the camps is absolute. Nor are airports ideal spaces for meetings or even handing out leaflets.

Despite these obstacles, there was a steady growth in numbers in the first half of 2008. Workers who lived in the Pilbara, not FIFO workers, provided the bulk of the first wave. For them, management style and workplace issues were not the only sources of discontent. The impact of the mining boom on local living conditions was just as much a problem.[27] This was strikingly so in Karratha where costs of living had gone sky high. With two-thirds of the rail workforce based there, Karratha became the union's heartland. Inland, the workers at Tom Price (the major base for mining operations) were more likely to be FIFO, and, initially, less likely to be unionised.[28]

The company itself put workplace concerns back on the agenda, with its announcement in mid-2008 of plans to introduce 'autonomous train operations' (commonly referred to as ATO) – that is, driverless trains – on its mainline tracks.[29] For some time, there had been rumours about what Rio's much touted 'Mine of the Future' might look like in practice. Here was a powerful hint – with the suggestion that these trains would not only be automated but that control of them would be shifted out of the Pilbara altogether. The drivers were not only alarmed at the threat to their jobs but, as we will see later, they were, as skilled workers have been since the first industrial revolution, affronted by their work being codified, measured and then, potentially, taken away.[30] Automation also brought into play what many had long seen as the injustice of Rio's unilateral power over its workforce. The union's *Newsletter* argued that collective action was needed because: 'If we are still on individual agreements at the time that ATO is introduced and the company decides that redundancies are necessary we will have no means by which to protect ourselves'.[31]

The automation announcement obviously concentrated people's minds after a year or so of more general discussions in the workforce about the rights and wrongs of unionism. Among other things, this led to a focus on getting clauses into any agreement with Rio that would guarantee consultation with the union. Still, the new unionists agreed with the union leadership that winning the collective agreement was the priority. In other words, the sheer fact of having a collective voice in the heart of Rio was of more concern than every detail of the agreement.[32] As for membership growth, this was not a goal in itself – it was not a matter of gloating about how many unionists they had – but using the numbers as the means to forcing the company to negotiate.[33]

Developing a set of agreed claims did involve moments of internal conflict. The more militant (or simply more frustrated) members wanted to make ambitious demands on the company backed with an aggressive industrial strategy. However, the leadership's view, which prevailed, was driven above all by that desire to 'get back into Rio'.[34] Increasingly, the push for the collective agreement was shaped by the threat of automation, especially seeking to limit managerial power through mandating a consultative committee. Other demands reflected long-standing complaints (which had surfaced at the start of the PMU campaign) over managerial prerogative, in particular over dispute-resolution procedures, the rights of union delegates to deal with individual grievances, time off for delegates to attend agreement negotiations, performance reviews, and changes to how 'book-offs' worked (when drivers had to stay away from home at the end of a shift).[35] Again, and in essence, control, not more money, was the overriding issue.

After the federal election of 2007, it was now the union, not the company, which sought to take advantage of the section of the law which allowed agreements to cover only a *part of* an enterprise. In this case, that part would be defined by work – the rail section – not by the time of appointment as had the PIEA. This plan simply reflected the fact that the CFMEU was at this stage the only body trying to unionise workers anywhere in Rio's iron ore operations. The concerns of one group of workers, potentially members of that one union, came together with the capacity of the union and a legislative opening to shape the initial success in membership growth among the rail workers.

The union's strategy was not, though, just fixed on the Pilbara itself. The federal and state officials of the union believed they needed a legal strategy to challenge Rio's control of the workplace. The struggle would be waged in both the crib-room and the courtroom.

The Pilbara and the courts

In June 2008, the CFMEU lodged a claim with the Australian Industrial Relations Commission for a collective agreement for the rail section of Rio.[36] More than three years later, in September 2011, the agreement was finalised. In this time, the union campaign harnessed workers' concerns about automation as membership grew. Furthermore, because a claim had been served on the employer, union members were able to take 'protected' (that is, lawful) industrial action in support of their demands. Getting an agreement was a complex and messy process, overlapping with other court cases, with the first strikes for a generation and the introduction of Labor's new laws. The reader needs to be as patient as the lawyers on both sides seemed to be. In summarising it here, it should be said that, as complicated as this story is, things seem neater on the page than they were for everyone involved at the time.

The company wanted to retain its preferred position of not bargaining with a union at all. It formally refused to bargain in August 2008.[37] In response, the union considered strike action only to discover one of the many legal obstacles which would come into play: that different entities actually employed the workers, not simply Rio Tinto. At law the employers were Pilbara Iron, which was a new overall company for Rio's Pilbara operations, and the two predecessor organisations, Hamersley Iron and Robe River.[38] Nonetheless, in September, the union won approval from the Commission to ballot its members over taking industrial action.[39]

The proposed strikes were an attempt not only to pressure the employer but to build the profile of unionism itself and, with that, the union's membership. In taking the first industrial action at Rio since the strike of 1992, the union was cautious because of fears among the union workforce that strikes would be tied by the company and the media to the worst fears of the Pilbara's 'bad old days'.

The union was also circumscribed by the extraordinarily detailed strike legislation in Australia, which is among the most complex in the world. In this case, only about forty workers – those who were not covered by AWAs – could take (lawful) strike action.[40] As the ballot was underway, a spokesperson for Rio spelt out the company's commitment to 'direct engagement', telling the union that the company did 'not intend to isolate sections of [the] workplace and treat them differently to others'. This meant that it would be 'inappropriate to enter union discussions'. Rio did exactly what the unionists had feared the company would do. It called up its reading of history: the Pilbara had been 'an industrial nightmare' in the past.[41] Nonetheless, the workers did vote to take strike action. On 11 October, they went out for twelve hours, and other stoppages followed. Union members were bolstered by the knowledge that other workers coming off expiring AWAs would be able to join them in undertaking lawful industrial action in the future.[42]

Thereafter, the campaign to re-unionise this section of Rio was carried on away from the public eye. The local Lodge built up numbers while federal union leaders (and Rio) lobbied the

Labor Party to reshape its legislation. For would-be unionists in the Pilbara, the exact shape of the new laws was a very significant issue. The major benefit was the likely introduction of 'Majority Support Determinations' (MSDs) which the union might use to force Rio to the bargaining table. However, because Labor had not gone back to old-style arbitration with its new laws, employers could probably stall negotiations. 'Good faith bargaining' clauses might not necessarily create better relationships at work or, as arbitration had done, enforce the resolution of disputes.[43]

That the complexities of labour law really mattered to – and were understood by – members of the union was quite clear. The Lodge informed workers that 'we should be able to get to the table with the company without the need for any more industrial action'.[44] This may have been optimistic but rightly or wrongly it led to a change in strategy, shifting the ground of the attack on Rio out of the workplace and back into the judicial and regulatory system.[45]

There were two parts to the legal strategy. The union decided to test the validity of the PIEA. If the agreement could be done away with, then the company's non-union plan would be threatened. The second step was to take advantage of the new laws by seeking an MSD to show that the union had support for a collective agreement.[46]

The legal challenge to the PIEA began in June 2009. The CFMEU asked the Federal Court to declare the agreement invalid, arguing that, although the Act allowed agreements to be made for 'part of a single business', the legislation had never contemplated that this apply on the basis of the date of the commencement of employment. To be less technical about it, the argument was that the company had been a bit too tricky; too clever by half. The Court reserved its judgment.

With this matter unresolved, the union pushed ahead with the other part of its strategy, applying to the new federal tribunal, then called Fair Work Australia, for an MSD.[47] This was the stage at which the impact of new federal legislation was most obvious in the Pilbara. It proved to be a turning point in the long battle with Rio. In January 2010, the company changed course. It agreed to bargain with the union without an MSD being tested. For union officials this was an acknowledgement by Rio of the support the union had garnered on the ground.[48] Rio's decision occasioned a flurry of media coverage, most of which seemed to imagine that this had happened overnight, solely because of the changed law. In fact, as we have seen, it came after a long campaign and could not be understood without realising that it was about local conditions and activities as well as what happened in parliament and courts.[49]

This was still only a beginning, not the end. A major break with the Pilbara's recent history it certainly was, but now the union had to get something from the bargaining process. For eighteen months on and off from February 2010, company and union representatives met. The union probably gave more ground than it would have in other sites to try to secure that all-important sign-off to make the collective agreement operative.[50]

Many of the claims that members desperately wanted were equally important as points of resistance and principle to Rio, including retaining company control over annual pay increases and performance bonuses, and resisting formal recognition of the union itself. Rio would not accept a consultative committee to which union – as opposed to 'employee' – representatives would be elected.[51] Most of these matters fell the company's way. The agreement was not concluded until June 2011. It was then endorsed by the workers in August and approved by the tribunal in September.[52]

In the meantime, the overall dispute between Rio and the CFMEU had become even more complex. The union's case against the PIEA failed when the court ruled that the agreement was perfectly valid, but the CFMEU decided to appeal to a full bench of the same court. Meantime, its latest strike plans were now thwarted by the PIEA itself. The company successfully opposed the union's request to hold a strike ballot. Because the agreement sought to cover some workers employed under the PIEA, the company argued that it was not valid. Fair Work Australia agreed. It declared in October 2010 that: 'a protected action ballot order cannot be sought [where] some employees who would be covered by the proposed enterprise agreement are covered by a "non-expired: in the first instance" agreement-based collective transitional instrument'. What that meant was the PIEA.[53] The company and its lawyers were delighted. The union's state secretary, Gary Wood, was not. He argued that the rights of a majority of employees were still being thwarted – after all the legislative changes that had taken place – by non-union arrangements entered into under the previous government's laws.[54]

Within a few weeks of the union's rail agreement being finalised, the CFMEU won a major victory when a Full Bench of the Federal Court declared the PIEA invalid. Once again, there were delays as Rio sought leave to appeal to the High Court.[55] This move was rejected in February 2012; the invalidity of the PIEA was confirmed.[56] No doubt the threat to direct engagement lay behind the decision made by Rio to throw such time and money into these cases.

A small group of unionised workers in the booming Pilbara appeared to be making significant advances after twenty years, the PMU aside, of near union invisibility. At this very time, though, the ground shifted under them. In the same month that the PIEA was buried, Rio announced more-specific plans for rail automation now under the label of 'Autohaul'.[57] As the next chapter will show, the company would now try to change the nature of the work these men and women did – and even where they did it.

The boom, politics and the unions

For all its remoteness, the Pilbara lay at the epicentre of a truly spectacular mining boom, one which tied the Australian economy ever more closely to global markets. Supposedly overpaid workers, the FIFO phenomenon and the El Dorado–like wealth of the sector became regular media fodder. More importantly, the boom revealed deep contradictions in Australia's politics and economics. The boom was built on what we might have thought were outmoded assumptions about family: the typically male FIFO worker, and with him the industry itself, relied on stay-at-home families and partners often in low-wage regions.[58] These networks supported itinerant workers and delivered them safe and sound to the airport, and the company, ahead of each swing in the Pilbara. At the base of the boom was a kind of family life more in line with a nineteenth-century settler society than a twenty-first century world.

Australia in the boom was both newly wealthy and deeply dependant, both an urban society and China's quarry. The country had policy-makers who could find neither the wit nor the will to harness the boom. As for the power of the companies: they brought a federal government to heel just as they had long bent state governments to their will. In many ways, the boom was not new but a throwback to speculative eras in the nineteenth century. The roots of this lay in the dominance which the companies had established over work and workers in the Pilbara itself with their fightback from 1986 on.

The boom did not affect employment relations as might once have been expected, strengthening the unions' hand. The employers had remade the Pilbara in nearly twenty years of conflict before this latest boom; it was no longer a union space. The upheavals elsewhere in the country over labour law and policy affected how the companies played the employment relations game but did not of themselves bring the unions back into that game. The rail sections at BHP Billiton and then Rio were the only collectively represented sections of the Pilbara's iron ore sites. Was the key question in all this why these workers were unionised? Or why so few others were? Either way, the union's leader, Gary Wood, explained their success in Rio in terms of 'the Fair Work Act and the determination of the rank and file to stand united in support of their collective rights'.[59] The legal context, which had been so vital in the national political debate, shaped union opportunities. The law was a necessary, but not sufficient, element to changing the Pilbara, as the continued failure of unions to re-organise the rest of the Pilbara made clear. The union had some credibility with rail workers; but, across the spectrum, employer dominance remained.

Chapter Ten

Beyond the Boom

MINING HAS ALWAYS been a cyclical business. Its greater globalisation, the growing complexity of production networks and the short-term focus of major investors have made that unevenness more marked than ever before. The mining firms have always tried to insulate themselves from global pressures and local politics to keep their businesses profitable. It might be said that all firms in global markets face these pressures, but what is different about mining is its geography. Mining firms are constrained by the fixity of deposits as well as by external factors. As we have seen time and again, the companies have done all they can to overcome these constraints, reworking the human geography of the Pilbara to enhance their sway over the place.

From 2012, the Pilbara iron ore employers faced a major challenge. At the height of the hype about the boom, everything seemed to change. Having weathered the global economic crisis of 2008–09, they were hit by a fresh problem. A mere slowdown of the Chinese rate of growth – bear in mind its economy was still expanding – was enough to send mining's export earnings, employment numbers and share prices down.

In the Pilbara, the downturn played out in complex ways. The companies had committed themselves to major expansions.

Bigger ports and new mines were under development and the big three, Rio Tinto, BHP Billiton and FMG, had for a time been signalling their intention to raise tonnages. For the two older companies, iron ore had been a source of wealth in otherwise uncertain times; for FMG, with higher costs, falling prices could be especially troublesome. Here we briefly examine how all of this unfolded before going back to look at what it all meant for the longer-term vision that Rio had for automation in mining.

The global downturn and the Pilbara

The slowing of China's growth, and with that its demand for steel, was enough to bring down the iron ore price very markedly. No-one could have seriously believed that prices of more than US$150 a tonne were sustainable over the longer term – and if investors had, then they were likely to be in trouble – but the downturn had major effects in and beyond the Pilbara.

Apart from a brief rally in the second half of 2012, the iron ore price fell from early 2011, markedly and consistently from 2013. The impact of these changes on the companies was potentially greater than ever because most sales of ore were now (and had long been) on 'spot prices'. In the early days, prices had been set in advance. Back then, only Robe River, with its lower-quality ores, had sold at spot prices, and we have seen what that had meant for employment relations in their sites. Selling at spot prices was all right in the boom, but it added massive pressure in more uncertain times. The ore price bounced around a bit, but the trend was clear enough. It dropped to about $100 a tonne in the second half of 2013, recovered a little, then fell in most months through to early 2016 when it sold for nearer $50 a tonne. Cost pressures on most new, smaller companies were intense. They had come into production late in the boom and typically had massive debts. At prices anything like as low as $50, they would struggle. Some suspended their operations and some closed down; amid growing uncertainty, Atlas Iron suspended work at its three Pilbara mines in April 2015.[1]

Between the smaller operators (some of them in other parts of the west, outside the Pilbara) and the big three companies lay two high-profile people, Gina Rinehart and Clive Palmer. Already extremely wealthy, both were keen to develop their mines – Rinehart to do what her father had not done and run her own mine, and Palmer, well, because he could.[2] Palmer had immense political influence in Queensland and would soon form his own political party, but for now his enthusiasm to get into iron ore led to an unhappy alliance with Chinese investors. Ore was exported from a new site, Cape Preston near Karratha, just as ore prices began to fall. The commercial partners ended up in a long-running court case by which time Palmer was more concerned with political troubles and the collapse of a nickel refinery he had run in the east. In some ways, Palmer was a nineteenth-century figure, a flamboyant individual and high-profile speculator as opposed to a serious iron ore mining capitalist.

Rinehart was a more important player. This status derived from her inherited fortune and, like many other entrepreneurs in the west, her larger-than-life profile. Her family feuds were probably better known than her views on mining employment relations and politics, but these became clear enough when she joined the anti-tax campaign. One of her firms, Hancock Prospecting, had, on one account, paid 'an effective tax cash rate of 10.6 per cent' between 2003 and 2008.[3] Well after the mining tax furore had died down, it was revealed that she and her companies were not alone. Transnational resource corporations operating in Australia were often paying well below what should have been the standard, if they were paying any tax at all.

Back in the Pilbara, Rinehart did have employment relations concerns in setting up her operation at the Roy Hill mine and seeking visa exemptions, claiming she could not secure the necessary unskilled labour.[4] This created a political storm, as did her remarks that Australians should understand that 'Africans want to work, and…are willing to work for less than $2 per day'. At the same time she sought to revive an old dream of recasting not just the Pilbara but northern Australia in general as a special economic zone 'with fewer regulations and taxes, a region that truly welcomes investment and people'.[5] One might have thought that someone whose wealth was based on pastoralism, inheritance and royalties might have been a little more temperate in providing counsel to the nation. Despite the problems she felt her company faced, the $10 billion project exported its first shipment from Port Hedland in December 2015.[6]

The major drivers and key players in the Pilbara were still Rio Tinto and BHP Billiton, not least because of how they chose to handle falling prices. Their response to the fall in prices was at first sight puzzling and certainly controversial: they increased their production. The critics, not least Andrew Forrest and others at FMG, argued that this move only pushed the price down further and that the two biggest companies were using the downturn to drive smaller rivals to the wall. Forrest went so far as to call for a cap on exports and won some political support for the idea, although nothing came of it. The two big companies sailed on. They were estimated to have their costs down to a remarkably low $20 a tonne, so that even with ore prices at less than $50 they were still more than covering costs.[7]

Measured by volume, then, the downturn did not really register. Rio's exports actually rose from about 200 to 263 million tonnes as prices fell between 2013 and 2015, while BHP Billiton's output went up from about 159 to 218 million tonnes. Both companies, however, experienced reductions in earnings, a downside they seemed willing to absorb for enhanced market share.[8] They also had their eyes on the longer term, with the downturn allowing them to work on maintenance of their rail networks over 2016 and 2017. Despite its criticisms of the big two, FMG had come through intact by 2016 with its share price rising fast and with the prospect of further sales if BHP Billiton and Rio did lose time with planned maintenance and development work.[9]

Where unions, wages and conditions had been the focus in previous downturns, the companies now focussed more on changes to rosters (especially at FMG) and in finding savings in their dealings not just with labour but with other firms. BHP Billiton and Rio did cut jobs both on mine sites and in their city offices, but they were also, and perhaps mainly, engaged in reshaping their inter-business operations.

The impact of the downturn reverberated from China, through the two biggest companies into the Pilbara and across the state. Mining engineering companies found that they were not receiving orders for new equipment, and maintenance was being put off. Costs were much more closely monitored and old commercial relationships undone as jobs were put out to tender. Even companies that were global players themselves, such as Caterpillar's local suppliers and Komatsu, found themselves struggling. The impact of the downturn was even greater in other mining areas such as coal, and on smaller regional communities such as the Hunter Valley and towns in north Queensland, but in Perth there were also job losses, shorter hours, changes in work organisation and general uncertainty across the suburbs where the engineering companies' employees worked and lived.[10] By April 2016, pressures had become so intense that media and even the state government bit back at Rio. Like BHP Billiton, the company had doubled its own payment times to suppliers, leading to a wave of complaints from those smaller companies.[11]

Despite the downturn, analysts remained optimistic about the future of iron ore mining, forecasting that Australia's share of global output would increase.[12] The cost of producing iron ore in Australia remained low by global standards, while reports of the end of the boom in China were, most sober judges agreed, rather exaggerated.[13]

Mine automation

One curious aspect of all these global changes and their local echoes was the way in which they intertwined with politics and unionism in just one part of one company: Rio's rail network. No sooner had workers in the rail section won a union agreement than the company committed US$518 million to train automation. And just as that happened, ore prices began to fall. Where did these overlapping changes leave the automation planning – and how would they affect those newly unionised rail workers? In the pits, there had long been trials of automated trucks, while train loading and unloading and ship loading had, of course, been automated for some time.[14] Rail automation, however, was a new frontier.

Rio's original automation announcement in February 2012 had said little about employment relations beyond asserting that because of company growth and increases in production there would be no job losses, and asserting that managers would 'engage directly with those affected'.[15] Yet the change was, potentially, truly radical, not just another round of technological change, but also a geographical one and a social one. It spoke to wider uncertainties about the future of work across the country. The job of driving the main-line trains would shift not only out of the locomotive cabin in the Pilbara, but to a computer centre in Perth. This refixing of the industry's geography could see train controllers working in an office environment more like an IT or a finance workplace than a traditional mining site. The mining industry would be coming to the metropolis. Trains would be 'driven' by a new workforce made up of employees living in the suburbs, not the mix of FIFO and Pilbara workers which, against the odds, the union had recently organised.

As with inland mine sites anywhere, Rio's rail network is central to the overall production system. It has long been a remarkable technical achievement: the ports at Dampier and Cape Lambert are up to 400 kilometres from Rio's fourteen scattered mines. Every day and night of the year, trains are driven between mines and ports, up to eighteen trips to the port every twenty-four hours.[16]

The company's CEO, Sam Walsh, explained the thinking behind the move in telling a national television program that the change was 'all about mining more efficiently, more safely, more productively and creating a competitive advantage'.[17] The press release had said much the same, locating automation within the company's global 'Mine of the Future' program. It was all about reducing costs, increasing efficiency and improving 'health, safety and environmental performance'.[18] Greg Lilleyman, Rio's President Pilbara Operations, told an interviewer that, in effect, mining would be simpler without human beings: '[w]e can plan ahead and schedule with much greater certainty with an automated system, than you can with the inherent issues that come along with people being involved'.[19] As we have just seen, some groups of workers could indeed be more troublesome to the company than others.

On the same television program, a financial analyst was more direct about cost-cutting. Accommodation and flight costs would be cut and the company could say to itself: 'instead of having someone we have to pay $300,000 a year, say, to be on-site…all the extra cost of fly in/fly out, we can get a skilled person for $75,000 or $80,000 a year to do that same job'.[20] Cutting costs would be, then, a result of the proposed geographical changes, that is, changes to where the work would be done. This was, potentially, a startling transformation in company strategy. After years of increasing reliance on FIFO workers, costs would be cut in one (unionised) part of the operation by shifting the site of work itself. Work would now come to the worker, not the worker to the work.

Nothing was said by the company's leadership or the media about the recent re-unionisation of the rail section in Rio, but it is not hard to make such a connection. Ever since the unions had been removed in the early 1990s, Rio had been sure that (weather aside) it could sustain the continuous production flows vital to shipping, customers and revenue. Twelve-hour shifts and rosters had also smoothed out the process. With automation, every step in the labour process from extraction to ship-loading had become more integrated, a process which had begun in the 1970s and was now, it seemed, moving rapidly ahead. The words of Rio's CEO summed it all up: 'my iron ore business is a logistics business'.[21]

Most read this kind of statement, rightly, to be positioning mining as a more sophisticated industry than one which simply dug up dirt. The CEO was also getting at something else: a unionised workforce at the link between pit and port was a threat to the whole system. For Walsh, unions were still an unwanted outside force, not a legitimate form of voice for the company's workers. Late in 2011, when the CFMEU was making progress in rail, Walsh, then head of the company's iron ore division, had made this plain. He told a forum 'that a union agenda can be different to what our organisation, what our people…needs'.[22] That you could not be in a union and be 'our people' was clear.

If automation meant a shift to high-end technologies and, in this case, a spatial transformation of work, then this would be a shock to workers and a threat to their union. To assess these changes from that viewpoint, we can go back to the workplace itself, starting in the main rail depot at Karratha. For the drivers based in Karratha, the starting point for any twelve-hour shift was the Seven Mile Depot just outside the town. These workers were, as we have seen, not typical Pilbara iron ore workers because they were more likely than not to begin the work day by getting out of their own bed at home, not setting out from a FIFO camp.[23] They were also more likely than not to be union members. The Depot's crib room spoke of the importance of these changes. As well as company material about automation on the walls and notice-boards, there were information sheets about town events but also, for the first time in years, union news. If the working day began in domestic 'normalcy', global resonances were never far away: the drivers pointed out that their locomotives were built in the USA and the ore cars in China, with the steel made from Australian iron ore. Not many local jobs to show for all those Pilbara ores.[24]

The locomotive cabin itself bore the marks of the changes to come, with data-storage devices recording and codifying the drivers' skills, and screens for braking systems and train information systems. Throughout the journey, despite being out on their own across vast distances for much of the shift, drivers were very aware that their work was being watched and controlled. A Track Asset Protection system monitored train performance with a screen displaying the train's axle count (in effect, the number of cars attached), along with data on temperatures and electrics. The cabin was tied into the company's network, filled with radio conversations between the train controllers and other drivers.

On most shifts, the drivers had no advance knowledge of what their work would be like, how far they might be driving, if they were driving at all. The train controllers made decisions on those factors in accord with the complex needs of organising the trains that moved between loaders at the mine and dumpers at the port. All of these arrangements needed to slot into the worker's twelve-hour shift, and individual drivers needed to be close to their depot at the end of each shift. Train or track failures, weather problems, or changes to staff levels could unsettle any planning – and it was some of these uncertainties that automation was designed to address.

For many outside the Pilbara, not only were the drivers overpaid, they were also doing a job that was not very demanding. As craft workers always have, most drivers saw things differently. The job was about tacit knowledge and discretionary skills. These were things they still prized despite the rise of technology. This came through in any explanation of the work. Drivers talked about not just driving the train but 'feeling' it. A total of 30,000 tonnes of steel and ore might not seem to be a flexible thing, but drivers spoke of them in that way. Despite (or because) of the trains being nearly 2.5 kilometres long, they had a lot of give in them, with sixteen metres of slack when at a standstill. Being so long, a train underway across the Pilbara could easily be on different topographies at any one moment, its front end climbing, middle on the flat, the rear end still coming down a slope. So, drivers talked about the job in terms of art, not just science: 'you always need to know where your last car is, and drive to that... you want the train to squeeze up a bit and the back to drive through the dips'. One summed it up almost poetically: the task was to get to your destination without 'tormenting the train'.[25]

This particular struggle for control of (a small part of) the Pilbara's industry was the latest, and very clear, playing out of older and wider struggles between the imperatives of managers' control or, as they saw it, rationality, on one hand, and worker discretion and cultures on the other. The company's manuals, which prescribed how to handle the train through different sections of tracks, lay at the heart of one view of work. Nobody denied their usefulness, but to insist that there was only one way to do the job offended drivers. The response – that 'they take away a lot of job pride when they over-regulate; they dumb it down'[26] – could have been taken from many different moments of conflict over fifty years of mining the Pilbara.

For the company's leaders and managers, the rail maps and charts were merely technical and topographical; they were objective documents which defined routes and instructed drivers. In regarding them this way, they defined the Pilbara as a mining site alone and, within that, as a company space. The maps were not totally unimaginative – there were rail-sidings named after wildlife such as Ibis, Koala, Lizard, Lyre, Pelican and Possum, to name a few.

It makes for compelling listening when workers explain that they understood what was happening to them in explicitly geographical terms. To them, the rail system was as much about their own lives, local social activity and sense of place as it was about mining and production, or, as their CEO put it, logistics. For experienced drivers, especially local ones, the maps had other meanings. It was not just that some of them thought that the maps and charts were an insult to their skills – after all, they could see in their minds' eye every bend, hill and dip on the 1,500 kilometres of Rio's Pilbara train tracks – they also read the maps differently from how they were drawn. They made them into their own kinds of places. They could see and point to where they had camped with children, fished with family, prospected with friends.[27] The Pilbara and the maps were not just geographies of production work but, for some, geographies of life.

In short, if much of the story of proposed automation was an old tale about control, skill and job security, then its geography gave it a twist. The automation systems could have been located and staffed in Karratha. The potential shift to faraway Perth gave Rio the possibility of making major cost savings by bringing the work to the workforce, turning the logic of FIFO on its head and perhaps presaging other changes. As one driver asked me rather rhetorically: 'if they can drive the trains from Perth, then why not from Hyderabad?' That was a concern by no means unique to the Pilbara.

Unions in a new Pilbara?

What these changes in technology and geography mean for employment relations and unionism remains to be seen, because the process is still unfolding and indeed stalling in 2017. Making sure that there was a consultative committee to which union representatives might be elected had been a key element of negotiations for the rail section's collective agreement. Unionists had decided to settle for the term 'employee representatives' instead of 'union representatives' in the agreement, but, as it turned out, union members secured a majority of the committee's positions in May 2012.[28] This did not mean that arguments about the process were finished. The company's notices about automation informed workers that committee meetings would not deal with the 'process for managing the impact on individual employees'.[29]

In the following four years, the union moved as near it could to normalising its place in the company. It was not going to disappear in a hurry.

It may well have been the case that, with the automation proposal, Rio and the union had locked themselves into a tricky spiral: the company, in pressing ahead with automation plans, drove more workers into the arms of the union, but by virtue of its own success in unionising those workers, the CFMEU faced the reality that Rio might now be still more committed to pressing on with automating and relocating the work away from those troublesome Pilbara workers. More specifically, delays to the implementation added to uncertainty and, in some quarters, scepticism about the change. This did no harm to the union's claim that an independent and collective voice was a good thing for those working in the rail section.

Four years after the rail automation announcement, Rio had made much less progress than had been envisaged. By March 2016, it was clear that the program would not only be delayed but would cost more than had been planned. One reason for this was the logistical complexity of the rail system; another was the general slowdown in the resources sector – a change once again driven by global product markets. By the middle of 2016, the company was openly talking about having underestimated the 'mechanical work' involved while, behind the scenes, at least some of the rail managers were soft-pedalling on the issue and indeed on their previous hostility to the CFMEU. Meantime, the other operators were in no hurry to make similar changes.[30]

The CFMEU's plan to concede contentious points where reasonable to get a deal with Rio had reflected the symbolic importance of making any inroads at all into so harsh a territory for union organisation. It had paid off, with the union hanging in there and, in 2015 and 2016, beginning to examine broadening its coverage in rail and making a new agreement.

Any further advance in total numbers of union members would require more than one union to be active and, despite the problems of the past, it would likely necessitate cooperation between them. The CFMEU was largely confined to rail and the AMWU and ETU to their respective groups of skilled workers, fitters and electricians respectively. This left the question of what role the union with the biggest potential coverage, the AWU, might play. That there might be scope for cooperation after the PMU catastrophe spoke to the passage of time and also, outside the Pilbara, of a rapprochement between the leadership of these two big unions. As it happened, the west had been the site of a seemingly unlikely alliance between the AWU and another union with which it had long had political differences, the Maritime Union of Australia. Since 2003, the unions had been working more closely in the offshore oil and gas sector. If there were local examples of cooperation, there were also wider realities: the unions had worked together in international forums and the leaders had established cordial relations.[31]

So it was that in May 2013, after many months of discussions, the federal leaderships of the CFMEU and AWU announced that they had come to an agreement to work together towards making wider inroads into Rio's Pilbara sites. The intention was that new members would join the nascent Western Mine Workers' Alliance, with the two unions sharing numbers between them. For the first time for many years, full-time union organisers – one from each union – were placed in the Pilbara, based in the bigger inland town of Tom Price. Overall, the strategy owed something to the PMU in that it aimed to simplify union access for workers to whom the complexities of union coverage might seem arcane. With just two unions involved and a concentration on just one employer, albeit the Pilbara's biggest, the plan was a little simpler this time around.[32]

This latest attempt at unionising in Rio also took place in a changed environment. Since the collapse of the PMU a decade earlier, the sheer fact that the Rio rail section had remained unionised had modified (if not basically altered) the Pilbara's workplace dynamics. It was one thing to say that it was undesirable to have unions back in the Pilbara. It was another thing to say it was impossible. It had actually happened. The company seemed to be more open to drivers' concerns and to negotiation about their conditions. For their part, the drivers were more amenable to 'work-throughs', that is, taking trains the full distance from Tom Price to the coast. This provided a considerable cost-saving to the company, something which had earlier been behind the push to automate the trains. Union leaders detected a more tolerant attitude from managers in rail (under a new GM, Zara Fisher) and in senior HR based in the west. Apparently reconciling themselves to the fact that the 80 per cent of the rail section who belonged to the CFMEU were not going away, the company's orientation seemed to be changing. Even the talk of direct engagement was played down.[33]

As had been the case a decade earlier, the combined unions' organising strategy – as in many other places around the world – was low-key. Although many yearned for a more militant campaign and a wider political struggle against the mining giants, the strategy was a cautious one. It was based on the implicit assumption that a new kind of message was needed for these relatively well-paid but non-union workers who had all come of age in a world utterly transformed since the 1960s and 1970s. With the politics of the labour movement being so different overall from the radicalism of those days, unionism in any one place was likely to start off in a much more defensive manner.

There was also a broader trend to deal with: changes in public opinion. A view that blue-collar workers being well-paid was somehow improper seemed to date back to the debates surrounding the national waterfront dispute of 1988, and was certainly alive and well in the west. The few remaining unionised workers in other resource sectors, most notably the offshore gas industry, were routinely pilloried as some kind of irresponsible 'other', dangerous to the future of the economy. A group of mineworkers in the Pilbara could not do much to unsettle these ways of thinking, but these wider attitudes – reminiscent of the way that even fellow unionists had dubbed the train drivers as 'koala bears' back in the 1970s – no doubt were another factor that shaped how they went about their union business.

The Western Mine Workers' Alliance focussed its publicity around its attempt 'to make your workplace fairer' and promised that a 'professional organisation' would speak for members and provide representation and legal advice if needed. At union meetings, formalities were kept to a minimum and, as in other recent initiatives in community and union groups, the leaders urged attendees themselves to drive the agenda with issues that mattered to them and in ways that suited them.[34]

A new union venture the Alliance might have been, but many of the issues were at least as old as the PMU, and indeed almost as old as the industry itself. In a word, they were about community. Even FIFO workers were coming to share the locals' concerns that company responses to the downturn were undercutting community conditions and services still more. Workers urged the company to commit to residents' jobs instead of the seemingly default position of using Perth-based contractors. Contrary to common perceptions, these men and women felt that full-time residential workers were being 'let go' due to the downturn; and that it was not, by any means, the contractors alone who were absorbing the cost pressures. Meantime, they complained, more and more houses in Rio's two inland towns were empty; shops were closing; medical services were under threat; even the pub in Tom Price had reduced its opening hours. As had long been the case in the Pilbara, they also felt acutely the pressure that many faced when their children reached high-school age, there being only one high school for the two inland towns. Parents had to choose between this option, sending the kids down south to board, quitting the job or going FIFO. If this was a tough personal decision, it was also one with union implications, because the loss of older workers was a loss of priceless experience from the union memory bank.[35]

By the middle of 2016, the Alliance had enrolled several hundred members working at Rio's two oldest mines, in Paraburdoo and Tom Price, and on the pits outside Paraburdoo on the Eastern Ranges. The total membership was still small in the scheme of things but noteworthy in three ways: 60 per cent were FIFO workers; about one in four were women (a slightly higher proportion than the numbers in the mine workforces); and a concentration in one area, the crane workers, allowed the workers to push for a second collective agreement, alongside rail's. The FIFO base was not in the same proportion as Rio's workforce, 80 per cent of whom were now estimated to be FIFO, but it was a marked improvement on previous attempts to organise these workers. One reason was simply that as the downturn hit, one of the outcomes was cost-cutting around the edges of FIFO conditions. The unions also worked hard to break down the divisions between FIFO and locals by referring to the former as 'resident FIFOs'.

As to the prospect of a new union-based agreement in Rio, the best hope for the pro-union workers lay with the crane section. There were fewer than 100 workers doing this job but it was a vital one, not least when emergencies arose on the mines or railways. This work could not easily be contracted out, and the workforce was therefore well placed to insist on its rights. Most importantly, it was also possible for the union to argue that the crane section was an 'operationally distinct' part of the organisation and therefore, like the train drivers, eligible under the law to have its own agreement. By late 2016, moves were underway to draw up a set of demands which would form the basis for negotiations with the company for just such an agreement.[36]

In passing, it should be said that it is more than a touch ironic that occupations, if not old-fashioned craft work itself, should be the basis for re-unionisation all these years after the old demarcations were said to be irrelevant. So much media, academic and union research was (in many sectors rightly enough) focussing on the volatility of careers and the fluidity of jobs to say that specific occupational identification and interests were no longer important. That was not so, it appeared, in all sections of the Pilbara.

A new geography of work and mining

Even with the problems that had befallen rail automation, the proposed changes remained something of a lightning rod for more general discussion of mining in and after the boom. That was understandable because the idea that an industry so supposedly 'old economy' as open-cut mining could be the site of significant technological changes and, with them, geographical changes fascinated many people.

The attention given to a long-term trend such as automation and, on the other hand, to weekly fluctuations in ore prices, hid from view the changes that had already taken place in the Pilbara. These were developments which had, all but literally, taken the ground cut from under organised labour's feet and posed fundamental questions about work in the industry. Because of the growth of FIFO labour, mineworkers were, by the middle of the 2010s, far more likely to live in Perth or other urban centres than in the Pilbara. They were increasingly likely to be working for manufacturers, assemblers and maintenance companies in the metropolitan suburbs rather than as support workers in the mining towns. In the ten years after 1996, as the boom had picked up, one analysis made plain the extraordinary change in mining's geography: employment in the industry had more than doubled in Perth but had only risen by 29 per cent in the state's regions.[37] More than that, FIFO had reshaped not only place but time. The rhythms of work life and also social and family life now revolved around twelve-hour shifts and the particular sorts of rosters under which the workforce was working.[38]

As the mining boom in the Pilbara slowed and then became a downturn, the reactions were different from earlier periods in which ore prices had fallen. This was not simply because ore prices in the boom had been so excessive and the falls so rapid, but also because the Pilbara had changed so dramatically in the last twenty years. Production and contracting networks meant that the companies' responses were now focussed on new forms of cost-cutting. With a generally non-union workforce, the Pilbara was no longer so obviously a site of conflict when market conditions changed. Media attention and popular discussion concentrated on the impact on smaller mining companies and suppliers, along with the impact on government revenues. Yet the fundamental issues examined in this book – namely, work, employment relations and the politics around them – were still important. Because production was actually ramped up by the major companies, the remaining unions were in a position to make some small advances, something not always associated with the workplace politics of a downturn. As had been the case since the 1960s, local agitation emerged over community and workplace issues alike. Familiar questions and conflicts remained vital across the much-changed Pilbara.

Part Five:
From the Deserts Profits Come

Chapter Eleven

Remaking the Pilbara

THE PILBARA ONCE excited hope, and some dread, about workers and their power. It seemed, drawing on the words in A. D. Hope's poem *Australia*, that from the deserts prophets might come, prophets of a different kind of unionism, winning for workers not just high wages but control of their work and towns. It was not to be: if anything, the companies were the prophets. They were prophets of new forms of employment relations, which reshaped not just the desert but the cities.

As exports of iron ore were first shipped from the Pilbara in the mid-1960s, mining employers developed the first of their 'spatial fixes' to solve two fundamental geographical problems: the physical remoteness of the place and the fixity of the ores themselves. Their operations were literally grounded in the fabulous wealth on and below the Pilbara's surface. These major, global companies were bound by place, and by a difficult place at that.

Untapped for aeons, the iron ore in the Pilbara was rendered valuable by global steel-making and made viable by global capital. A place long understood as isolated and remote, long defined purely by its physical geography, now became 'tethered to the world'. It was discovered afresh as if those mining it were willing away the idea that anybody had been there before the pastoralists. Forty-thousand years of human life were forgotten; the Indigenous people and their rights to lands were even more marginalised than they had been under pastoralism.

During the first twenty years of mining, the companies' solutions to the problems posed by mining the Pilbara (their first 'spatial fix)' were hemmed in by accepted work norms and labour laws, and by the nature of the workers and their aggressive unions. At first the companies accepted unions and built towns. It is not too much to say that in these years, workers and their unions and communities fixed the Pilbara as their kind of place, driving a series of improvements in wages and conditions and winning an almost unparalleled say over many aspects of their daily work. None of this came easily and, with each downturn, the employers tried to shift the frontier of control back in their favour. Nonetheless, local behaviours and images confirmed both in the local mind and the national imagination that there was a Pilbara way of doing things – a union way.

The explanation for the success of this kind of union power is neither solely geographical, that is, because of the Pilbara's physical isolation, nor due simply to the nature of the work and workers. Rather it was both; it was driven by the workers' orientation to 'their' place. Friendship, black humour, a certain kind of male ethos and also family networks lay beneath the power that workers exercised.

From the late 1970s, as the global economy was transformed and employers in most countries reworked their attitudes to unions and collective bargaining, every one of the factors that had shaped the Pilbara iron ore industry was transformed. With these changes to global and local politics, the mining companies were more able to fight back against unions. They acted in a less reactive and episodic way and with more coherent and sustained plans than in the past.

So it was that after internal change in some of the companies and local skirmishes with their unions, the Pilbara's tensions burst onto a public stage with the lock-out of the workforce at Robe River in 1986. Over the next thirteen years, the workplace power (if not the wages) that the unions had achieved was wound back, and the unions themselves were almost wiped out as, one by one, the three big operators, Robe River, Hamersley Iron and BHP, went on anti-union offensives. In doing so, the companies reworked not just their mine sites but the very nature of the Pilbara. Company strategy backed by changes to labour law reshaped local work and employment relations and laid the foundations for a new human geography – a different community life – in the Pilbara. To say this is not to romanticise life before those changes, but to acknowledge that FIFO and longer working hours did transform life in the Pilbara. FIFO weakened communities and neutralised them as sources of union power. Not one mining town was built after the 1970s despite massive expansion of the mines.

One of the great mysteries of the Pilbara's iron ore history is why unionism collapsed so quickly, especially at Hamersley Iron. Was its strength more apparent than real? Perhaps it was, but to make that claim does not really answer the question: it needs an explanation itself. What emerges from this book is something at first sight counter-intuitive: that perhaps the unions had not been radical and united enough. Of course, in the 1960s and 1970s, the workers were militant in that they were unafraid of a stoush and they managed to negotiate major changes in wages and conditions. Even in those days, though, whenever they were divided among themselves or when they, and their place, the Pilbara, could be portrayed as somehow wild and 'out there', then they had less power than seemed to be the case. As some observers said at the time, only a handful of workers and a few of their leaders sensed that the companies and governments might one day come to take them down. On the other hand, the unions, if united and in touch with their towns, had the potential to remake not just wages and conditions but unionism and community itself in the Pilbara and for others; but the chance that this could all happen was eroded over time.

Since those days, through the biggest boom in the Pilbara's history, the mining corporations have sustained their redefinition of the place. The Pilbara is in many ways now less a place to live than it is a globally fixed place of production. It is now hardly even accurate to speak of the 'Pilbara iron ore industry': it is both more and less than that. It is more than that because it is more globalised than ever before in terms of capital, with traders and speculators more and more influential. It is also susceptible to the mere slowing in the rate of growth in one country, China. At the same time, though, it is 'less than the Pilbara' because of local changes. In a sense, the industry has shifted out of the Pilbara. FIFO labour, city-based maintenance and remote operation centres have seen to that. Mining families and maintenance workers live in metropolitan communities, and white-collar operators work in urban computer centres.

During the boom, because of this relocation of the workforce, there was still no revival of towns, let alone unionism. Sure enough, then and since, there have been challenges to this non-union work arrangement, with an attempt at locally based unionism through the PMU, the persistence of unionism among the rail workers and the work of the new Western Mine Workers' Alliance. There have been political challenges to the companies, too, with changes to labour law and new tax proposals, but the companies worked around one while seeing off the other. As a federal election took place in 2016, the mining lobby groups ramped up the pressure with detailed proposals for still more legislative change to strengthen their control over labour.

By 2017, the Pilbara was a very different place from the 1970s. It was no longer, as some workers had hoped, a site for prophets from unions or communities; rather, it was about profits for companies and shareholders. Amid all this change, the focus on material benefits, on profits and wages, points rather ironically to continuities in the Pilbara story. When individual contracts were introduced from the 1990s, they had to match union wages to win the workforce over. In a capital-intensive industry such as iron ore, the struggle to remove union power was about control over production, not earnings. So, to this day, the Pilbara's wage structures are still marked by the union past, as much as managers and many workers may like to think otherwise.

Today, the dominant storyline about the Pilbara is about what has been gained: a story of colossal tonnages and fabulous wealth. Yet there is another take on all this. At the outset, I made the perhaps too obvious point that without labour – without work being organised and workers being on the job, be it on the mine or in a remote centre – the ores stay in the ground. Putting work and workers at the heart of the Pilbara iron ore industry means that we have a very different kind of history. It is a story of loss as much as gain. Others have explained this loss in terms of the Indigenous past and present. So it is with work and employment relations. The militant unionists could not find a path between global corporations' anti-unionism on one side and, on the other, the national forces at first of 'consensus' and, after that, government anti-union laws. As they struggled to hold onto the Pilbara from the 1980s, the landscape around them was remade.

The defeats of the Pilbara unions between 1986 and 1999 made it much easier for the mining corporations to make other changes. They made fundamental changes to the very geography of Pilbara life, through FIFO labour, twelve-hour shifts and the weakening of community. With the unions and the Pilbara brought under control, these global mining firms went on to influence national policy-making and labour markets, and jobs that were very different from the Pilbara's. State governments no longer thought of their earlier commitments to downstream development and local content in big projects. National governments would run scared of tax intervention and wealth funds.

It is often said, and rightly so, that work and employment relations have been wholly transformed in all sectors in the last generation – and that more change and less certainty is coming. It is also the case, sadly, that how working people dealt with the places and times in which they found themselves in the past is either written out of our memory or held up to ridicule. All of this has very markedly been the case with the Pilbara. The assault on its particular kind of unionism has been matched by an assault on the memory and knowledge of it. This book has tried to do something to fix that, to rethink that past and to see the present differently.

The supposedly old-fashioned world of strikes and unions was not simply critical to the daily activities of managers and workers in the Pilbara for over a generation, as important as all that was. The result of those conflicts was enduring: a reshaping of the Pilbara and national politics. What happened in the Pilbara was central to corporate and government ambitions elsewhere, to reworking what unionism means to people across the country and to how we think about the role the global resources sector might play in contemporary Australia. From the deserts came not only profits but power across the Australian terrain.

Notes

Notes to Chapter Two – 'Tethered to the World'

1   In *The Roaring Days*, published in 1889, Lawson laments the passing of a boom and the taming of the bush but, specific mining cycles aside, the image is just right for the Pilbara's economic geography.

2   Western Australia's Government Geologist quoted in N. Dufty, *Industrial Relations in the Pilbara Iron Ore Industry,* Western Australian Institute of Technology, Perth, 1984, p. 5; see also G. Blainey, *The Rush That Never Ended: A History of Australian Mining*, Melbourne University Press, Carlton, 2nd edition, 1969, p. 350.

3   Blainey, *The Rush That Never Ended*, p. 350; Dufty, *Industrial Relations in the Pilbara*, p. 5.

4   See Blainey, *The Rush That Never Ended*, p. 350, for the bands 'like the stripes of a zebra'.

5   Dufty, *Industrial Relations in the Pilbara*, p. 27.

6   R. M. Berndt, 'Traditional Aboriginal Life in Western Australia: as it was and is', in R. M. and C. H. Berndt (eds), *Aborigines of the West: Their Past and Their Present*, University of Western Australia Press, Nedlands, 2nd edition, 1980, pp. 7–8. For languages, see Wangka Maya Pilbara Aboriginal Language Centre at http://www.wangkamaya.org.au/pilbara-languages/information-on-pilbaras-languages (accessed 15 January 2016).

7   R. M. Berndt 'Traditional Aboriginal Life in Western Australia: as it was and is'; on the hangings in NSW, R. Milliss, *Waterloo Creek: The Australia Day Massacre of 1838, George Gipps and the British Conquest of New South Wales*, McPhee Gribble, Ringwood, 1992.

8   ibid, p. 152. More generally on early white expansion, see G. Bolton, *Land of Vision and Mirage: Western Australia Since 1826*, University of Western Australia Press, Crawley, 2008, pp 33–5.

9   See, among others, the foundational work of C. D. Rowley in *The Remote Aborigines*, ANU Press, Canberra, 1971 and *The Destruction of Aboriginal Society*, Penguin, Ringwood, 1972. On the Pilbara itself, J. Wilson, 'The Pilbara Aboriginal social movement: An outline of its background and significance', in R. M. and C. H. Berndt (eds), *Aborigines of the West*, pp. 151–5; J. Peck, 'Excavating the Pilbara: A Polanyian exploration', *Geographical Research*, 51(3), 2013, pp. 229–34.

10 On the Martu past and present, see M. Lewis, *Conversations with the Mob*, University of Western Australia Press, Crawley, 2008.

11 Bolton, *Land of Vision and Mirage*, p. 34; N. Olive, *Enough is Enough: A History of the Pilbara Mob*, Fremantle Arts Centre Press, Fremantle, 2007, pp. 68–71.

12 Olive, *Enough is Enough*, especially pp. 59–66.

13 ibid, chapters 2 & 3. For a more sanguine view of pearling, see J. Hardie, *Nor'Westers of the Pilbara Breed*, Shire of Port Hedland, Port Hedland, 1981, pp. 26–7.

14 McLeod quoted in D. Wilson, *Different White People: Radical Activism for Aboriginal Rights 1946-1972*, UWA Publishing, Crawley, 2015, p. 58.

15 M. Hess, 'Black and Red: The Pilbara pastoral workers' strike, 1946', *Aboriginal History* 18(1–2), 1994, pp. 65–86; B. Oliver, *Unity is Strength: A History of the Australian Labor Party and the Trades and Labor Council in Western Australia*, API Network, Perth, 2003, pp. 175–7; M. Hale (Minyjun), *Kurlumarniny: We Come From the Desert*, Aboriginal Studies Press, Canberra, 2012, chapters 2 & 3.

16 O. White, *Under the Iron Rainbow: Northwest Australia Today*, Heinemann, Melbourne, 1969, p. 25.

17 Blainey, *The Rush That Never Ended*, p. 345.

18 G. S. Bain and R. Price, *Profiles of Union Growth: A Comparative Statistical Portrait of Eight Countries*, Basil Blackwell, Oxford, 1980, pp. 121–3.

19 R. Fells, 'Award restructuring – but where did we get our awards in the first place? The development of the award and union structures in the iron ore industry', in C. Fox (ed.), *Papers in Labour History 11*, Australian Society for the Study of Labour History, Perth, 1993, p. 4.

20 ibid, pp. 5–6.

21 R. Lockwood, *War on the Waterfront: Menzies, Japan and Pig-iron Dispute*, Hale & Iremonger, Sydney, 1987.

22 Blainey, *The Rush That Never Ended*, pp. 345–8; M. Griffiths, *Of Mines and Men: Australia's 20th Century Mining Miracle, 1945–1985*, Kangaroo Press, East Roseville, 1998, pp. 56–7, 64–6.

23 A. H. Panton quoted in Blainey, *The Rush That Never Ended*, p. 347.

24 Blainey, *The Rush That Never Ended*, p. 348.

25 ibid; on BHP see especially K. Tsokhas, *Beyond Dependence: Companies, Labour Processes and Australian Mining*, Oxford University Press, Melbourne, 1986, pp. 96–7.

26 Fells, 'Award restructuring', pp. 3–4.

27 M. Knox, *Boom: The Underground History of Australia, From Gold Rush to GFC*, Viking, Melbourne, 2013, p. 308; for the whole saga, see pp. 306–14. For the original debunking, N. Phillipson, *Man of Iron*, Wren Books, Melbourne, 1974, chapter 7.

28 Griffiths, *Of Mines and Men*, pp. 57–8.

29   Knox, *Boom*, p. 308. The details of the complex corporate negotiations to set up the mines are beyond the scope of this book but there are good accounts in Griffiths, *Of Mines and Men*, chapter 5; A. Trengrove, *Adventure in Iron: Hamersley's First Decade*, Stockwell Press, Mont Albert, 1976, chapters 4–8; Phillipson, *Man of Iron*, chapters 8-13; Tsokhas, *Beyond Dependence*, pp. 66–71, 96–105.

30   Knox, *Boom*. p. 313.

31   ibid, pp. 311–13.

32   J. Horsley, 'Conceptualising the state, governance and development in a semi-peripheral resource economy: the evolution of State Agreements in Western Australia', *Australian Geographer*, 44(3), 2013, pp. 291–4.

33   Blainey, *The Rush That Never Ended*, pp. 350–51.

34   Griffiths, *Of Mines and Men*, pp. 58–9.

35   Blainey, *The Rush That Never Ended*, p. 349.

36   Charles Court in 1970 quoted in H. Thompson, 'The pyramid of power: transnational corporations in the Pilbara', in E. L. Wheelwright and K. Buckley (eds), *Essays in the Political Economy of Australian Capitalism: Volume Five*, Australia and New Zealand Book Company, Sydney, 1983, p. 80.

37   Thompson, 'The pyramid of power', pp. 80–81.

38   J. Tracey, 'The construction phase of the Pilbara iron ore industry 1965–1972: Workers, their unions, and organising the industry', *Papers in Labour History*, 13, Australian Society for the Study of Labour History, Perth, 1994, p. 16.

39   This argument is made by several authors in M. Brueckner, A. Durey, R. Mayes and C. Pforr (eds), *Resource Curse or Cure: On the Sustainability of Development in Western Australia*, Springer, Heidelberg, 2014, building on E. J. Harman and B. W Head (eds), *State, Capital and Resources in the North and West of Australia*, University of Western Australia Press, Nedlands, 1982.

40   M. Coyne and L. Edwards, *The Oz Factor: Who's Doing What in Australia*, Dove Communications. East Malvern, 1980, p. 68.

41   For intriguing hints about this, see Hardie, *Nor'Westers of the Pilbara Breed*, pp. 234–5, 242, 247–8.

42   To be precise, in the ten years after 1961; see Tracey, 'The construction phase of the Pilbara iron ore industry'. For male population figures, Dufty, *Industrial Relations in the Pilbara*, pp. 35–40.

43   See Griffiths, *Of Mines and Men*, p. 62, on the 'utter madness' of this.

44   H. Thompson, 'The capital-labour relation in the mining sector', *Journal of Australian Political Economy*, 15, 1983, pp. 64–85, on Robe River; for the others see Tracey, 'The construction phase of the Pilbara iron ore industry'; H. Court, 'Industrial Disputes in the Pilbara', unpublished MEc thesis, University of Western Australia, 1976, p. 25; Trengrove, *Adventure in Iron*, chapters 6 & 7; Griffiths, *Of Mines and Men*, pp. 58–60.

45   Tracey, 'The construction phase of the Pilbara iron ore industry', p. 19; see also M. Bickerton, *Dust Over the Pilbara*, Artlook, Perth, 1980, p. 96.

46   Tracey, 'The construction phase of the Pilbara iron ore industry', pp. 19, 21.

47　Utah's Vincent Kontny quoted in J. Hardie, *Nor'Westers of the Pilbara Breed*, p. 243.

48　Tracey, 'The construction phase of the Pilbara iron ore industry', p. 19.

49　ibid, pp. 19–20; Fells, 'Award restructuring', pp. 9–11.

50　Employers' Federation's John Ince quoted in Fells, 'Award restructuring', p. 8.

51　Fells, 'Award restructuring', pp. 9–12.

52　Tracey, 'The construction phase of the Pilbara iron ore industry', p. 22; S. Reid, 'Gil Barr's story, Part 2', in C. Fox and M. Hess (eds), *Papers in Labour History 5*, Australian Society for the Study of Labour History, Perth, 1990, especially pp. 19–20 (family), 22–24 (union rivalry).

53　Tracey, 'The construction phase of the Pilbara iron ore industry', p. 20; Reid, 'Gil Barr's story, Part 2', p. 20.

54　The strongest recent statement of the wider importance of coal-mining unionism can be found in T. Mitchell, *Carbon Democracy: Political Power in the Age of Oil*, Verso, London, 2011, chapter 1. Good accounts of coal mining, community and politics are too numerous to list here but, for a particularly subtle analysis, see A. Metcalf, *For Freedom and Dignity: Historical Agency and Class Structures in the Coalfields of NSW*, Allen & Unwin, North Sydney, 1988.

Notes to Chapter Three – Frontiers of Control

1　E. P. Thompson, *The Making of the English Working Class*, Penguin Books, Harmondsworth, 1975 (first published 1963), p. 13.

2　There are surprisingly few accounts of work itself in the very first years of the industry. For overviews: Trengrove, *Adventure in Iron*, chapter 11; R. Fells and S. Reid, '"Tons were the go": Some recollections of work in the early days of the Pilbara iron ore industry', in C. Fox and M. Hess (eds), *Papers in Labour History 8*, Australian Society for the Study of Labour History, Perth, 1991, pp. 1–16.

3　Iron Ore History Project, interview with Finucane, pp. 12–13. These interviews were undertaken by Ray Fells and Stuart Reid for a University of Western Australia oral history project, 1990–91. Most are available from the Battye Library. I am grateful to Ray and also to Rob Lambert for providing me with original copies of the transcripts.

4　Tsokhas, *Beyond Dependence*, pp. 155–8, records the increasing technical sophistication of the mines. 'Trial and error' is his phrase.

5　C. L. Goodrich, *The Frontier of Control: A Study in British Workplace Politics* (1920); Foreword by R. H. Tawney. Republished with a new foreword and additional notes by R. Hyman, Pluto Press, London, 1975. The conception/ execution dichotomy comes from H. Braverman, *Labor and Monopoly Capital*, Monthly Review Press, New York, 1974.

6　For the comparison, see Tsokhas, *Beyond Dependence*, pp. 149–54. See also E. Eklund, *Mining Towns: Making a Living, Making a Life*, New South Publishing, Sydney, 2012; B. Ellem and J. Shields, 'Making a "union town":

Class, gender and workers' control in inter-war Broken Hill', *Labour History*, 78, 2000, pp. 116–40.

7   Fells and Reid, '"Tons were the go"', p. 6. This summary is taken from their interviews, pp. 4–11.

8   On the rail system, see Trengrove, *Adventure in Iron*, chapter 12.

9   Tsokhas, *Beyond Dependence*, pp. 154, 160; C. Bulbeck, 'The iron ore stockpile and dispute activity in the Pilbara', *Journal of Industrial Relations*, 25(4), 1983, pp. 431–44.

10 For an acute summary of these divisions, see Thompson, 'The pyramid of power', p. 85.

11 ibid.

12 Dufty, *Industrial Relations in the Pilbara*, pp. 77–84. The state's train drivers' union showed little interest in iron ore because of the small numbers of workers so far away from the south of the state; see Fells, 'Award restructuring', pp. 13–16.

13 Reid, 'Gil Barr's story, Part 2', p. 22.

14 See Thompson, 'The pyramid of power', pp. 84–8, for an insightful argument about managerial control and division between workers.

15 Iron Ore History Project, interview with John Beales, p. 83.

16 Fells and Reid, '"Tons were the go"', p. 13.

17 Thompson, *The Making of the English Working Class*, p. 9.

18 Court, 'Industrial Disputes in the Pilbara'; Fells, 'Award restructuring'; H. Thompson, 'Class and gender in the Pilbara', *Arena*, 68, 1984, p. 124.

19 B. Dabscheck and J. Niland, *Industrial Relations in Australia*, George Allen & Unwin, North Sydney, 1985 (first published 1981), pp. 273–4.

20 Tracey, 'The construction phase of the Pilbara iron ore industry', p. 21; Fells, 'Award restructuring', pp. 7–12; E. P. Lovett, 'The Structure of Industrial Negotiations within the Pilbara Iron Ore Industry', unpublished MA thesis, Monash University, 1980, p. 70.

21 Fells, 'Award restructuring', p. 13, on turnover; Lovett, 'The Structure of Industrial Negotiations within the Pilbara Iron Ore Industry', pp. 36–7, on Hamersley more broadly.

22 Owen Salmon of the ETU (and later an eminent Commissioner himself) for the unions; quoted in Dufty, *Industrial Relations in the Pilbara*, p. 122. On the case more generally, see Fells, 'Award restructuring', pp. 12–17; Lovett, 'The Structure of Industrial Negotiations within the Pilbara Iron Ore Industry', pp. 70–74.

23 Most elegantly described by Doreen Massey as an 'impossible dichotomy' between the social and the spatial in *Spatial Divisions of Labour: Social Structures and the Geography of Production*, Macmillan, London, 1995 (first published 1984), p. 4.

24 Commissioner Kelly explicitly referred to these provisions: Lovett, 'The Structure of Industrial Negotiations within the Pilbara Iron Ore Industry', p. 73.

25 Dufty, *Industrial Relations in the Pilbara*, p. 123.

26 ibid for the union claim.

27 Lovett, 'The Structure of Industrial Negotiations within the Pilbara Iron Ore Industry', p. 35.

28 I am grateful to Ray Fells for suggesting this important line of argument to me.

29 Quoted in Dufty, *Industrial Relations in the Pilbara*, p. 123.

30 Reid, 'Gil Barr's story, Part 2', p. 28.

31 Lovett, 'The Structure of Industrial Negotiations within the Pilbara Iron Ore Industry', pp. 76–7.

32 ibid, p. 78.

33 Ray Fells reminded me of this classic formulation; see A. Flanders, *Management and Unions: The Theory and Reform of Industrial Relations*, Faber and Faber, London, 1970, p. 172.

34 H. Thompson, 'The Pilbara iron ore industry: Mining cycles and capital-labour relations', *Journal of Australian Political Economy*, 21, May 1987, p. 70.

35 Lovett, 'The Structure of Industrial Negotiations within the Pilbara Iron Ore Industry', pp. 77–8; S. Frenkel, 'Industrial conflict, workplace characteristics and accommodation structures in the Pilbara iron ore industry', *Journal of Industrial Relations*, 20(4), 1978, p. 400.

36 Correspondence with the author, 5 June 2016.

37 Lovett, 'The Structure of Industrial Negotiations within the Pilbara Iron Ore Industry', pp. 75–6.

38 ibid, p. 75; Dufty, *Industrial Relations in the Pilbara*, pp. 123–6, on grievance procedures and stewards.

39 Lovett provides an excellent critique: 'The Structure of Industrial Negotiations within the Pilbara Iron Ore Industry', pp. 78–83. See also Dufty, *Industrial Relations in the Pilbara*, pp. 183–4, for a clear summary of the end point.

40 These are views from Hamersley Iron and a report in the *National Times* as discussed in Bulbeck, 'The iron ore stockpile', p. 432.

41 Hardie, *Nor'Westers of the Pilbara Breed*, p. 247.

42 These extracts and stories are from Fells and Reid, '"Tons were the go"', pp. 2–3.

43 Iron Ore History Project, interview with Roger Parsons, p. 2.

44 Fells and Reid, '"Tons were the go"', p. 3.

45 ibid, p. 4.

46 ibid, pp. 4–5: John Buttery and Sid Elgar quoted.

47 Reid, 'Gil Barr's Story, Part 2', p. 27; Thompson, 'The Pilbara iron ore industry', p. 71.

48 Fells and Reid, '"Tons were the go"', p. 5: John Buttery and Barry Peters quoted.

49 ibid, p. 7: John Purdue quoted.

50 Fells and Reid, '"Tons were the go"', p. 14, quoting Bob Dickson.

51 ibid, quoting Barry Peters.
52 Frenkel, 'Industrial conflict', p. 401.
53 Fells and Reid, '"Tons were the go"', p. 11; see Nat Hilton on workgroup peer pressure p. 12.
54 ibid, p. 11.
55 Iron Ore History Project, interview with John Bryant, p. 5.
56 ibid, pp. 6 (backhoe), 9 (workshop).
57 ibid, p. 12.
58 Iron Ore History Project, interview with Roger Parsons, p. 12.
59 ibid, p. 13.
60 Reid, 'Gil Barr's story, Part 2', p. 21.
61 ibid, especially pp. 25–6.
62 ibid, p. 28.
63 Iron Ore History Project, interview with John Beales, p. 66.
64 ibid, interview with Jack Bainbridge, p. 3.
65 Kelly quoted in J. Read, *Marksy: The Life of Jack Marks*, Read Media, South Fremantle, 1998, p.144.
66 ibid, p.171.
67 Thompson, 'The Pilbara iron ore industry', p. 71; also Dufty, *Industrial Relations in the Pilbara*, pp. 73–4.
68 Iron Ore History Project, interview with Derek Miller, especially pp. 8–10. Miller was working for Goldsworthy and later became a key figure in BHP's shift to non-union agreements in 1999.
69 Reid, 'Gil Barr's story, Part 2', pp. 31–2.
70 Iron Ore History Project, interview with Ross Calnan, p. 3.
71 ibid, interview with Robert Dickson, pp. 8–10.
72 Dufty, *Industrial Relations in the Pilbara*, p. 63.
73 Thompson, 'Class and gender in the Pilbara', pp. 125–6.
74 Iron Ore History Project, interview with Ross Calnan, p. 24.
75 Thompson, 'Class and gender in the Pilbara', pp. 124–5.
76 Iron Ore History Project, interviews with un-named TWU Convenor, pp. 2–3; Archdale, CMEU Convenor, p. 28.
77 Tonkinson's work quoted in Peck, 'Excavating the Pilbara', p. 234.
78 H. Thompson, 'Normalisation: industrial relations and community control in the Pilbara', *The Australian Quarterly* 53(3), 1981, pp. 304–5.
79 As quoted in Tsokhas, *Beyond Dependence*, p. 159.
80 For a summary, see N. Dufty, 'Remoteness and industrial relations – the Pilbara iron ore industry', in T. B. Brealey et al. (eds), *Resource Communities: Settlement and Workforce Issues*, CSIRO Publishing, Melbourne, 1988, pp. 283–4.
81 Thompson, 'Normalisation', p. 305.
82 Iron Ore History Project, interview with Eric Archdale, CMEU Convenor, pp. 45–6.

83 ibid, p. 60.
84 Reid, 'Gil Barr's story, Part 2', pp. 29–30.
85 Thompson, 'Normalisation', p. 302.
86 Frenkel, 'Industrial conflict', p. 399.
87 Thompson, 'The Pilbara iron ore industry', p. 71.

Notes to Chapter Four – Contested Terrain

1   For an analysis of the academic arguments, see W. Brown, 'The Influence of Product Markets on Industrial Relations', in P. Blyton, N. Bacon, J. Fiorito and E. Heery (eds), *The Sage Handbook of Industrial Relations*, Sage, London, pp. 113–28.
2   This overview relies on the summary of the period in Thompson, 'The Pilbara iron ore industry', p. 71.
3   This phrase was first used in R. Edwards, *Contested Terrain: The Transformation of the Workplace in the Twentieth Century*, Basic Books, New York, 1979. It is a powerful way to think about control of work, especially when its geographical metaphor is taken more literally than Edwards intended.
4   Tsokhas, *Beyond Dependence*, pp. 155–8.
5   ibid, p. 156.
6   ibid.
7   Quoted, ibid, p. 157; spelling as in original.
8   Attested to in several of the Iron Ore History Project interviews, especially Finucane; un-named Goldsworthy Rail manager; Dains.
9   Tsokhas, *Beyond Dependence*, p. 158 on 'mindless' work; pp. 154–5 on work at Mt Newman more generally.
10 There is an excellent analysis of these wider issues in Thompson, 'The capital-labour relation in the mining sector'.
11 Dufty, *Industrial Relations in the Pilbara*, chapter 6, has the most detailed analysis.
12 The key source is Frenkel, 'Industrial conflict'; see also a summary in Tsokhas, *Beyond Dependence*, p. 154.
13 Tsokhas, *Beyond Dependence*, p. 155.
14 ibid, p. 290.
15 Dufty's summary in *Industrial Relations in the Pilbara*, p. 310.
16 Court, 'Industrial Disputes in the Pilbara', p. 161.
17 A point made in several interviews in the Iron Ore History project. See especially interview with Dains and others, pp. 34–5.
18 ibid, interview with Dick Keegan, FEDFU official, interview 1, pp. 5, 6, 10.
19 ibid, p. 96, on Goldsworthy.
20 This summary is taken from Dufty, *Industrial Relations in the Pilbara*, pp. 94–8.
21 C. Heath and C. Bulbeck, *Shadow of the Hill*, Fremantle Arts Centre Press, Fremantle, 1985.
22 ibid, pp. 39–43.

23 ibid, pp. 57–68.

24 ibid, p. 42.

25 ibid, p. 44.

26 Stewart Clements, FEDFU, interviewed: ibid, pp. 73, 75.

27 Dufty, *Industrial Relations in the Pilbara*, p. 36, for population data. On early recruitment practices, see W. Snell, The Formation of a Labour Force in a Developing Region: The Case of Hamersley Iron', unpublished MEc thesis, University of Western Australia, 1969.

28 See the critique in Thompson, 'Normalisation', especially pp. 305–21.

29 Quoted in Dufty, *Industrial Relations in the Pilbara*, p. 43.

30 ibid, pp. 42, 43, for these two quotations.

31 Thompson, 'Normalisation' is scathing about this failure.

32 D. Day, *papers*, n.d.: n.c.

33 This distinction is made in the work of Cathy Brigden, drawing on Richard Hyman, in 'Power and space in the Victorian Trades Hall Council'. In B. Ellem, R. Markey and J. Shields (eds), *Peak Unions in Australia: Origins, Purpose, Power, Agency*, The Federation Press, Leichhardt, 2004, pp. 219–35.

34 See Lovett, 'The Structure of Industrial Negotiations within the Pilbara Iron Ore Industry'; Dufty, *Industrial Relations in the Pilbara*.

35 Dufty, *Industrial Relations in the Pilbara*, p. 60, on AMMA in the 1970s.

36 This paragraph is based on Lovett, 'The Structure of Industrial Negotiations within the Pilbara Iron Ore Industry', pp. 84–7, and Dufty, *Industrial Relations in the Pilbara*, pp. 184–5.

37 Quoted in Dufty, *Industrial Relations in the Pilbara*, p. 185.

38 Lovett, 'The Structure of Industrial Negotiations within the Pilbara Iron Ore Industry', pp. 87–90.

39 ibid, p. 90.

40 ibid, pp. 91–2; see also Court, 'Industrial Disputes in the Pilbara'; Dufty, *Industrial Relations in the Pilbara*, pp. 187–8.

41 Lovett, 'The Structure of Industrial Negotiations within the Pilbara Iron Ore Industry', pp. 93–4; Dufty, *Industrial Relations in the Pilbara*, pp. 189–90.

42 See especially Lovett, 'The Structure of Industrial Negotiations within the Pilbara Iron Ore Industry', p. 94; Dufty, *Industrial Relations in the Pilbara*, pp. 188–9.

43 ibid; Dufty, *Industrial Relations in the Pilbara*, pp. 190–91; Frenkel, 'Industrial conflict', p. 400.

44 Quoted in Dufty, *Industrial Relations in the Pilbara*, p. 192.

45 Quoted in Lovett, 'The Structure of Industrial Negotiations within the Pilbara Iron Ore Industry', pp. 97, 98 respectively.

46 ibid, pp. 98–9, 101.

47 This account draws upon Dufty, *Industrial Relations in the Pilbara*, pp. 195–8; Lovett, 'The Structure of Industrial Negotiations within the Pilbara Iron Ore Industry', pp. 98–101.

48 Dufty, *Industrial Relations in the Pilbara*, p. 196.

49 Lovett, 'The Structure of Industrial Negotiations within the Pilbara Iron Ore Industry', p. 100.

50 Dufty, *Industrial Relations in the Pilbara*, p. 197.

51 ibid, p. 198.

52 ibid, pp. 217–26, for a very detailed assessment.

53 Based on the summary in Thompson, 'The Pilbara Iron Ore Industry', p. 68.

54 This account draws upon: N. Dufty, 'The 1979 Hamersley strike', *Australian Bulletin of Labour*, 8(4), 1982, pp. 210–28; Dufty, *Industrial Relations in the Pilbara*, pp. 201–15; Lovett, 'The Structure of Industrial Negotiations within the Pilbara Iron Ore Industry', pp. 101–14; H. Thompson and D. Bartlem, 'Confrontation in the Pilbara: The Hamersley iron ore strike', *Arena* 55, 1980, pp. 15–31. Read, *Marksy*, chapters 17–19 provides a good account of the role of the leading personalities.

55 Thompson and Bartlem, 'Confrontation in the Pilbara', p. 21.

56 Dufty, 'The 1979 Hamersley strike', p. 215.

57 This is also Lovett's view in 'The Structure of Industrial Negotiations within the Pilbara Iron Ore Industry', p. 105.

58 Reported in Thompson and Bartlem, 'Confrontation in the Pilbara', pp. 22–3.

59 ibid, p. 24.

60 ibid, p. 25.

61 Police Commissioner Leitch remained unapologetic when interviewed about 54B; see Read, *Marksy*, pp. 248–58. Among others on policing and the judiciary in the west at the time, see E. Blackburn, *Broken Lives*, Hardie Grant Books, South Yarra, 2001.

62 Thompson and Bartlem, 'Confrontation in the Pilbara', especially p. 29.

63 ibid, pp. 26–7; Dufty, 'The 1979 Hamersley strike', pp. 224–5.

64 Thompson and Bartlem, 'Confrontation in the Pilbara', pp. 27–9.

65 ibid, pp. 27–30, on the tensions around the outcome.

66 This summary is from Thompson, 'The Pilbara iron ore industry', pp. 72–4; see also several of the interviews in the Iron Ore History project, including Alf Kober.

67 Iron Ore History project, especially interview with Calnan, p. 25.

68 Thompson, 'The Pilbara iron ore industry', p. 73.

69 Iron Ore History project, interview with Calnan, p. 25; Parsons, pp. 25–7; unnamed TWU Convenor, p. 10; Dains and others, especially p. 43.

70 H. Thompson, 'Shorter working hours or redundancies: Class conflict in the Pilbara', *Arena* 56, 1980, pp. 113–27, for a full account.

71 ibid, pp. 120–25.

72 ibid, p. 125.

73 Quoted in P. Swain, *Strategic Choices: A Study of the Interaction of Industrial Relations and Corporate Strategy in the Pilbara Iron Ore Industry*, Curtin University of Technology, Perth, 1995, p. 213.

74   ibid, pp. 216–21 compared with Thompson's analysis, 'The capital-labour relation in the mining sector', pp. 71–3.
75   Swain, *Strategic Choices*, pp. 214–16.
76   Quoted, ibid, p. 224.
77   Quoted, ibid, p. 223.
78   ibid, pp. 225–6.
79   Tsokhas, *Beyond Dependence*, p.161.
80   Iron Ore History project, interviews with Beales, p. 71 – he claimed most did not even vote Labor; and with Mike Ireland, AMWU Convenor, pp. 2–3.
81   ibid, Beales, p. 74.

Notes to Chapter Five – War of Attrition: Robe River

1   There are still surprisingly few studies of the dispute. Swain, *Strategic Choices*, has a detailed chapter based on unrivalled access to management sources; my account is radically different but indebted to hers. The indispensable source for studies of the unions remains the work of Herb Thompson: H. Smith and H. Thompson, 'Industrial relations and the law: A case study of Robe River', *The Australian Quarterly*, 59 (3–4), 1987, pp. 297–305; Thompson, 'The Pilbara iron ore industry'; H. Thompson and H. Smith, 'Conflict at Robe River', *Arena*, 79 , 1987, pp. 76–91; on Thompson himself, see S. MacWilliam, 'Marginalizing opposition to the Accord', *Arena*, 92, 1990, pp. 146–50.
2   For a recent overview: C. F. Wright, 'The Prices and Incomes Accord: Its significance, impact and legacy', *Journal of Industrial Relations*, 56(2), 2014, pp. 264–72; also A. Scott, 'Looking to Sweden in order to reconstruct Australia', *Scandinavian Journal of History*, 34(3), 2009, pp. 330–52.
3   Swain, *Strategic Choices*, pp. 52–6.
4   Business Council of Australia, *Enterprise-Based Bargaining Units: A Better Way of Working*, BCA, Melbourne, 1989; for an overview of the period: B. Dabscheck, *Australian Industrial Relations in the 1980s*, Oxford University Press, South Melbourne, 1989.
5   H. R. Nicholls Society, *Arbitration in Contempt. The Proceedings of the Inaugural Seminar of the H. R. Nicholls Society Held in Melbourne 28 February – 2 March, 1986*, 1986, H. R. Nicholls Society, Melbourne.
6   Dabscheck, *Australian Industrial Relations in the 1980s*. For different reasons, the Labor Party government and the ACTU attacked apparently powerful unions, notably the Builders' Labourers' Federation in 1986 and, after the Robe River dispute, the Pilots' Federation.
7   Dufty, *Industrial Relations*, chapter 1; Swain, *Strategic Choices*, p. 50.
8   Dufty, *Industrial Relations*, pp. 9–10; Swain, *Strategic Choices*, pp. 66–9. Cleveland Cliffs' home operations were, and now as Cliffs Natural Resources remain, in the State of Ohio.
9   Smith and Thompson, 'Industrial Relations and the Law', p. 298; Swain, *Strategic Choices*, pp. 53–4.

10 Reported in W. Lowenstein, *Weevils at Work: What's Happening to Work in Australia*, Catalyst Press, Annandale, 1997, p. 179.

11 Swain, *Strategic Choices*, pp. 66–9.

12 C. Copeman, "The Robe River Affair" (H. R. Nicholls Society Archives, 1987), http://archive.hrnicholls.com.au/archives/vol3/vol3-8.php (accessed 5 November 2013); Swain, *Strategic Choices*, pp. 66–9; Thompson and Smith, 'Conflict at Robe River', pp. 77–8; Smith and Thompson, 'Industrial Relations and the Law', p. 298.

13 In the USA, Cleveland Cliffs' mines were unionised. T. Reynolds and V. Dawson, *Iron Will: Cleveland-Cliffs and the Mining of Iron Ore, 1847–2006*, Wayne State University Press, Detroit, 2011.

14 Even allowing for a degree of flattery – this was in a presentation to the Society – this is telling: Copeman, 'The Robe River Affair', no pagination: see last paragraph.

15 14 May 1986: cited among other places in P. Kelly, *The End of Certainty: The Story of the 1980s*, Allen & Unwin, St Leonards, 1992, p. 196.

16 Widely accepted anecdotally, this claim is sourced to a mining conference in the early 1990s in H. Thompson, 'The APPM dispute: The dinosaur and turtles vs the ACTU', *Economic and Labour Relations Review*, 3(2), 1992, p. 153.

17 Copeman, 'The Robe River Affair', no pagination.

18 ibid.

19 ibid; Swain, *Strategic Choices*, pp. 248–9; for a sympathetic explanation of the trigger to this, see P. Getlin, *The Power Switch at Robe*, Australian Institute for Public Policy, Perth, 1990.

20 Copeman, 'The Robe River Affair', no pagination.

21 ibid.

22 C. Copeman, 'I.R. Lessons from Recent Mining Industry History', H. R. Nicholls Society Archives, 1992, http://archive.hrnicholls.com.au/archives/vol13/vol13-7.php (accessed 5 November 2013).

23 Copeman, 'The Robe River Affair', no pagination.

24 Swain, *Strategic Choices*, pp. 243–50, has a full account of the lead-up to Peko's takeover.

25 ibid, p. 251.

26 Ian McRae, letter to, for example, the Australian Workers' Union, 31 July 1986, provided by Mr Bruce Bonnor (former Robe employee), copy in possession of the author; see also Swain, *Strategic Choices*, p. 236.

27 Ian McRae, *Memo To All Staff – Union Representatives*, 31 July 1986, memorandum provided by Mr Bruce Bonnor, copy in possession of the author.

28 Ian McRae, General Manager Operations, *Notice To All Employees*, 31 July 1986, memorandum provided by Mr Bruce Bonnor, copy in possession of the author.

29 Dabscheck, *Australian Industrial Relations*, p. 123.

30 Copeman, 'The Robe River Affair', no pagination; Swain, *Strategic Choices*, p. 247.

31 See especially Swain, *Strategic Choices*, p. 234.

32 Copeman quoted in Swain, *Strategic Choices*, p. 252.

33 These claims are still used against unions in the Pilbara: see for example the telling image of ice cream in a story in *Australian Mining Monthly*, September, 2011, p. 14.

34 Thompson and Smith, 'Conflict at Robe River', pp. 79–82; Swain, *Strategic Choices*, pp. 255–70 for a detailed account.

35 Swain, *Strategic Choices*, pp. 273–4; Thompson and Smith, 'Conflict at Robe River', pp. 79–80.

36 Thompson and Smith, 'Conflict at Robe River', pp. 82–3; Swain, *Strategic Choices*, pp. 270–71; author's interviews, Bruce Bonnor (former Robe River employee), Newman, 16 November 2005; Graeme Haynes (former ETU convenor), Perth, 23 November 2006.

37 Swain, *Strategic Choices*, p. 275.

38 Read, *Marksy*, pp. 321, 347–8; Thompson and Smith, 'Conflict at Robe River', pp. 80–81 for discussion of growing resentment; author's interviews, Bonnor; Haynes.

39 *Workplace News*, June 1989, no publisher details – copies in possession of author.

40 ibid.

41 Swain, *Strategic Choices*, p. 290; Read, *Marksy*, pp. 347–52; author's interviews, Bonner; Haynes; Jim Murie (ETU official), Paraburdoo, 16 November 2005; Lea Anderson (former Pilbara AMWU administrative worker), Perth, 27 November 2006.

42 Copies from Bruce Bonnor, covering 1987–88, in possession of the author.

43 Thompson and Smith, 'Conflict at Robe River', p. 84; Swain, *Strategic Choices*, pp. 284–8.

44 *Australian Financial Review*, 19 January 1987.

45 Thompson and Smith, 'Conflict at Robe River', p. 84.

46 Smith and Thompson, 'Industrial relations and the law', p. 302.

47 Swain, *Strategic Choices*, p. 290.

48 Thompson and Smith, 'Conflict at Robe River', p. 86.

49 *Australian Financial Review*, 23 December 1987, 5 January, 22 March 1988.

50 Tsokhas, *Beyond Dependence*, pp. 194–9.

51 Swain, *Strategic Choices*, p. 69.

52 Copeman, 'I.R. Lessons from Recent Mining Industry History', no pagination.

53 Thompson and Smith, 'Conflict at Robe River', especially p. 84; Read, *Marksy*, pp. 347–52.

54 Read, *Marksy*, pp. 348–50 cf Lowenstein, *Weevils*, pp. 178–80; author's interviews, Haynes and Anderson; various issues of *Resist to Exist* in author's possession. There is a set in the National Library.

55   Read, *Marksy*, p. 350.

56   Lowenstein, *Weevils*, p. 178.

57   Harold Peden in Lowenstein, *Weevils*, p. 179.

58   'Rob' in Lowenstein, *Weevils*, p. 180.

59   *Australian Financial Review*, 13, 22 April 1988.

60   AMSWU and others and Robe River Iron Associates, no. 50 of 1991, 73 *WAIG* 382.

61   Iron Ore History Project, interview with 'Railway Manager' Goldsworthy, p. 24; Derek Miller, p. 14.

62   *Australian Financial Review*, 5, 8 July 1991; for a full account see the Parliamentary statement by L. Graham, MLA, Pilbara, Western Australian Legislative Assembly, *Hansard*, 4 September 1991, pp. 4989–95.

63   H. Thompson, 'The ACTU and the struggle for irrelevance', *Green Left Weekly*, 7 April 1993, available at: https://www.greenleft.org.au/node/4315 (accessed 10 November 2014); see also *Australian Financial Review*, 10, 22 December 1992, and 4, 16 June 1993, 14 July 1994.

64   Various issues of *Resist to Exist* attest to this general hostility, as do author's interviews cited above; for some public hints, see *Australian Financial Review*, 1 December 1989; see also Lowenstein, *Weevils*, pp. 178–88; on writs, see *Australian Financial Review*, 13, 22 April 1988, and 1 December 1989; for Hamersley Iron, B. Hearn Mackinnon, *Behind WorkChoices: How One Company Changed Australia's Industrial Relations*, Heidelberg Press, Heidelberg, 2007, chapter 4.

65   Cited in *Australian Financial Review*, 1 December 1989.

66   Author's interviews with Dave McLane (former CMEU organiser), Perth, 23 November 2006; phone call with author, 20 June 2016.

67   AMSWU and others and Robe River Iron Associates, no. A4(1) of 1989, 69 *WAIG* 3000; AMSWU and others and Robe River Iron Associates, no. 2394, 1989, 70 *WAIG* 1659; AMSWU and others and Robe River Iron Associates, nos 3&4, 1990, 70 *WAIG* 2083.

68   AMSWU and others and Robe River Iron Associates, no. A4, 1987, 71 *WAIG* 582.

69   For legislative change in the Australian states, see S. Deery and R. Mitchell (eds), *Employment Relations: Individualisation and Union Exclusion*, The Federation Press, Annandale, 1999.

70   *Australian Financial Review*, 24 August 1993; for the cancellation of the award, see 73 *WAIG* 2646.

71   Author's interviews with Dave McLane, 23 November 2006; 20 June 2016.

72   Mackinnon, *Behind WorkChoices*, chapter 4.

73   *The Australian*, 31 December 1996.

74   Thompson, 'The APPM dispute'.

75   Thompson and Smith, 'Conflict at Robe River', p. 91.

76   Quoted in Swain, *Strategic Choices*, p. 260.

Notes to Chapter Six – Frontal Assault: Hamersley Iron

1   Unless otherwise referenced, this section is drawn from Mackinnon, *Behind WorkChoices*, chapter 1. I am deeply grateful to Bruce for sharing his expertise and contacts with me over many years.

2   Quoted, ibid, p. 6.

3   ibid, pp. 7, 8, for these quotations from 1981 and 1982.

4   J. T. Ludeke, *The Line in the Sand: The Long Road to Staff Employment in Comalco*, Wilkinson Books, Melbourne, 1996, p. 8.

5   ibid, pp. 4, 6, 7, for these elements of the strategy.

6   ibid, pp. 24–5.

7   ibid, p. 4.

8   Mackinnon, *Behind WorkChoices*, p. 3.

9   ibid, p. 9.

10 ibid, especially p. 30.

11 Mackinnon, *Behind WorkChoices*, chapter 3; Ludeke, *The Line in the Sand*.

12 Quoted in Mackinnon, *Behind WorkChoices*, p. 51.

13 This is from the company's 'Industrial Relations Plan', quoted in Swain, *Strategic Choices*, p. 227

14 Swain, *Strategic Choices*, p. 227; on the two-tier wage system, see B. Dabscheck, *Australian Industrial Relations in the 1980s*, Oxford University Press, Melbourne, 1989, chapters 5, 8.

15 Swain, *Strategic Choices* pp. 227–8; Thompson, 'The Pilbara Iron Ore Industry', pp. 76–7.

16 Iron Ore Production and Processing (Hamersley Iron Pty Limited) Award 1987, Division 1, Part 1: 1 Prevention of Disputes (copy in possession of author).

17 Swain, *Strategic Choices*, p. 228.

18 Iron Ore History Project, interview with Bob Dickson, interview 2, p. 10.

19 ibid, p. 11.

20 Swain, *Strategic Choices*, p. 229, quoting Hamersley's Industrial Relations Plan.

21 ibid, pp. 119–21.

22 ibid, p. 323.

23 This account is drawn from Swain, *Strategic Choices*, chapter 12 unless otherwise indicated.

24 ibid, pp. 331–2, quoting Commissioner Gregor.

25 ibid, p. 334, quoting Freeman.

26 This view is explained in Heath and Bulbeck, *Shadow of the Hill*; see also above, chapter 3.

27 Adapted from the summary in Swain, *Strategic Choices*, pp. 338–50.

28 These figures are from Mackinnon, *Behind WorkChoices*, pp. 55–7. This account is based on this source, chapter 4, unless otherwise indicated.

29 ibid, p. 58.

30 ibid, p. 60, quoting CRA Executive Karl Stewart.

31 H. Thompson, 'The APPM Dispute'.

32 Quoted in Mackinnon, *Behind WorkChoices*, p. 62.

33 R. Allen, 'Agreements without Unions', *Proceedings of the Conferences of the HR Nicholls Society*, vol. 15, 1994, available at: http://archive.hrnicholls.com. au/archives/vol15/vol15-7.php. For an earlier account of employment of non-unionists, see R. Allen, 'The end of the closed shop', *Proceedings of the Conferences of the HR Nicholls Society*, vol. 13, 1992, available at: http://archive. hrnicholls.com.au/archives/vol13/vol13-5.php.

34 HI Process Design Project, Implementation Manual, 1998. Copy in possession of author.

35 Mackinnon, *Behind WorkChoices*, p. 64.

36 This, of course, is a contested reading but this view was put to me forcefully in interviews with some of those on the ground at the time, most notably Dave McLane.

37 Allen, 'Agreements without Unions'; Mackinnon, *Behind WorkChoices*, p. 64.

38 Allen, 'Agreements without Unions'.

39 Mackinnon, *Behind WorkChoices*, p. 64.

40 I am grateful to ACTU Organiser Stewart Edward, who put me in contact with these two men and to the men themselves for telling me their stories. They did so in discussions in Karratha (Mark Troy, 6 December 2003) and Blacksmiths (Phil Randal, 11 August 2004).

41 This paragraph is based on a discussion with Steve Ward at his home in Tom Price in 2002. I am thankful for his time and for the collection of WAIC material which he gave me.

42 K. Heiler, R. Pickersgill and C. Briggs, *Working Time Arrangements in the Australian Mining Industry: Trends and Implications with Particular Reference to Occupational Health and Safety*, ILO, Geneva, 200, p. 18.

43 ibid, p. 13. These figures apply to metal mining generally.

44 ibid, p. 16. See also D. Peetz and G. Murray, '"You get really old, really quick": Involuntary long hours in the mining industry', *Journal of Industrial Relations*, 53(1), 2011, pp. 13–29.

45 From the company's submission to the Fair Work Review in 2012, available at: http://submissions.deewr.gov.au/sites/Submissions/FairWorkActReview/ Documents/RioTinto.pdf (accessed 2 May 2013).

46 http://www.riotinto.com/documents/The_way_we_work.pdf (accessed 8 March 2016).

47 State Secretary Gary Wood quoted in *Common Cause*, 'Pilbara mineworkers celebrate historic Rio breakthrough', 77(3), p. 7.

48 *Australian Financial Review,* cited in Allen, 'Agreements without Unions'.

Notes to Chapter Seven – 'The Last 500': BHP Billiton

1 For an overview of BHP: P. Thompson and R. Macklin, *The Big Fella: The Rise and Rise of BHP Billiton*, Random House, North Sydney, 2009.

2   *Employee Earnings, Benefits and Trade Union Membership, Australia*, Australian Bureau of Statistics, ABS 6310.0, various years; D. Peetz, 'Trend analysis of union membership', *Australian Journal of Labour Economics*, 8(1), 2005, pp. 1–24.
3   There are many overviews of this period. See, for example, R. Cooper and B. Ellem, 'Union power: space, structure, strategy', in M. Hearn and G. Michelson (eds), *Rethinking Work: Time, Space and Discourse*, Cambridge University Press, Cambridge, 2006, pp. 123–43.
4   J. Bailey and B. Hortsman, '"Life is full of choices": Industrial relations "reform" in Western Australia since 1993', paper for Association of Industrial Relations Academics of Australia and New Zealand, 14th Conference, Newcastle, February, 2000.
5   On the CFMEU and Rio Tinto, see Mackinnon, *Behind WorkChoices*.
6   ACTU, *unions@work*, Melbourne, ACTU, 1999; for an overview of the adoption of this strategy see R. Cooper, 'Organise, organise, organise! ACTU Congress 2000', *Journal of Industrial Relations*, 42(4), 2000, pp. 582–94 and for a later account by one of its architects, see M. Crosby, *Power at Work: Rebuilding the Australian Union Movement*, The Federation Press, Leichhardt, 2005.
7   Details of company strategy are taken from the subsequent Federal Court case: Australian Workers' Union v BHP Iron Ore Limited, Federal Court of Australia, 3, 2001, especially paragraphs 84–170.
8   ibid, paragraph 98.
9   ibid, paragraph 113.
10 For a summary of the origins of this see B. Ellem, 'Trade unionism in 1998', *Journal of Industrial Relations* 41(1), 1999, pp. 127–51.
11 *Australian Workers' Union v BHP Iron Ore Limited*, 3, 2001, paragraph 143.
12 Detailed in the above court case and in author's interviews with Will Tracey, Newman, 27 June 2001; Newman, 17 September 2001; Karratha, 19 September 2001.
13 Quoted in *Rock Solid* 1; Australian Workers' Union v BHP Iron Ore Limited, 3, 2001, paragraph 187.
14 This summary is drawn from: BHP Iron Ore, *Staff Handbook*; copies of rejected contracts kindly made available to me by union members; author's interviews, Tracey, 27 June 2001; Derek Schapper (Union counsel), Perth, 18 October 2001.
15 Author's interview, 'John', train driver, Port Hedland, 29 June, 2001; author's observation of delegates meetings, Newman and Port Hedland 26–29 June 2001.
16 *Australian Workers' Union v BHP Iron Ore Limited*, 3, 2001, paragraph 238; author's interview with Troy Burton (ACTU organiser), Sydney, 16 May 2002.
17 *Australian Workers' Union v BHP Iron Ore Limited*, 3, 2001, paragraph 237.
18 Author's interviews, 'John', train driver; author's observation of delegates meetings, Newman and Port Hedland.

19 *Australian Workers' Union v BHP Iron Ore Limited*, Federal Court of Australia, 430, 2000.

20 *Australian Workers' Union v BHP Iron Ore Limited*, 3, 2001, paragraph 190.

21 *Australian Workers' Union v BHP Iron Ore Limited*, 3, 2001.

22 H. Trinca and A. Davies, *Waterfront: The Battle That Changed Australia*, Doubleday, Milsons Point, 2000.

23 Author's interviews anonymous delegates, Newman and Port Hedland, 26–29 June, 17, 20 September 2001.

24 Author's interview, Burton.

25 ACTU, 'BHPIO – Port Hedland Campaign Planning Sessions', internal notes, 10–12 December 1999; notes made available to author; author's interview, Burton.

26 Biographical detail from Will Tracey; observations from nine research trips between 2001 and 2004 in particular.

27 Various issues of *Rock Solid*; author's interviews as above.

28 Author's interview, Ross Kumeroa (convenor), Newman, 26 June 2001.

29 *Rock Solid*; at this stage the newsletter was undated.

30 Author's interviews, Colleen Palmer (ASP), South Hedland, 28 June 2001; Maria Boyington (ASP), South Hedland, 29 June 2001.

31 Colleen Palmer, email with author, 19 August 2004.

32 ibid.

33 Author's interviews, Kumeroa; anonymous delegates, Newman and Port Hedland, 26–29 June, 17, 20 September 2001.

34 WAIRC, 04082 BHP Iron Ore Award, Western Australian Industrial Relations Commission 2001; WAIRC, 05810 AFMEPKIUA *vs* BHPIO & others, Western Australian Industrial Relations Commission 2002 for further reasons and decisions.

35 K. Storey, 'Fly-in/fly-out and fly-over: Mining and regional development in Western Australia', *Australian Geographer*, 32(2), 2001, pp. 133–48; see also Heiler, et al., *Working Time Arrangements in the Australian Mining Industry*; Peetz and Murray, '"You get really old, really quick"'.

Notes to Chapter Eight – The Rise and Fall of the Pilbara Mineworkers' Union

1 M. Bachelard, 'The day a myth called happiness died', *Weekend Australian*, 6–7 April 2002.

2 *Rock Solid*, 33.

3 J. Ford, 'Unions in the Pilbara', Parliament of Western Australia, Legislative Council, 22 August 2001. Copy kindly made available by Jon Ford.

4 Author's interviews, Tracey, Dampier, 27 June 2002; Stewart Edward, Paraburdoo, 16 November 2002.

5 ibid; also author's focus groups with Hamersley Iron workers in Paraburdoo, Tom Price, Dampier, Karratha, 24–29 June, 2002. Much of background in

this chapter is drawn from these focus groups while preparing an ACTU-commissioned report on the future of unions after the LK ballot.

6   ibid; Hamersley Iron (n.d.), 'Conditions of Employment'; documents kindly made available to author by anonymous employees.

7   Author's focus groups; interviews, Edward; anonymous small business owners, Paraburdoo, 16 November 2002.

8   Author's interviews, anonymous Hamersley employees, Tom Price, 18 September 2001; author's focus groups.

9   ibid.

10   Author's focus groups.

11   ibid.

12   Hamersley Iron, 'Employee Relations Feedback', 2003; report in possession of the author.

13   ibid; author's interview, Edward; T. Treadgold, 'Out of town and out of touch', *Business Review Weekly*, 27 June to 3 July 2002, p. 63.

14   The WorkChoices laws later confirmed this.

15   Pilbara Mineworkers' Union, Membership form, 2002, copy kindly made available by Stewart Edward.

16   Author's interviews, Tracey, en route, Tom Price to Dampier, 27 June 2002; Dampier, 27 June 2002; Karratha, 28 June 2002; Burton, Paraburdoo, 15 November 2002; Edward.

17   Author's interviews, Burton, 15 November 2002; Edward.

18   *Anvil*, 1, and in various issues thereafter.

19   Author's observations; interviews, Edward; Paraburdoo business owners.

20   S. Edward, personal communication with author, 17 April 2003.

21   Author's interviews, Burton, 15 November 2002; Edward; Kathleen Galvin (ACTU organiser), Paraburdoo, 15 November 2002; observation, Paraburdoo blitz, 15 November 2002.

22   *Anvil* 8-11; Edward, personal communication.

23   M. Priest, 'Union war threat after Rio deal', *Australian Financial Review*, 23 July 2003.

24   Australian Workers' Union, *Information and Documents Concerning the Federal Award Developments at Hamersley Iron & Robe River*, AWU, Melbourne, 2003.

25   S. Balogh, 'Back in business', *Weekend Australian*, 26–27 July 2003; author's observation, PMU meetings, Paraburdoo, Karratha, June 2004.

26   Author's observation, PMU meetings, Paraburdoo, Karratha, 23–25 June 2004.

27   Vic Davis, personal correspondence, 14 August 2004.

28   *Workforce*, 22 August 2003.

29   Attested to in various anonymous interviews.

30   AIRC, *Australian Workers Union and Hamersley Iron; Construction Forestry, Mining and Energy Union & others, Decision*, AIRC, Sydney, 2004, especially par. 50 and Appendix A.

31 Author's observation, PMU meeting, Paraburdoo, 24 June 2004; author's interviews, Tracey, Paraburdoo 23 June 2004; Karratha, 26 June 2004.

32 Author's observation, PMU meetings, Tom Price, Paraburdoo, Karratha, June 2004; author's interviews, Tracey, Karratha, 26 June 2004; Gary Wood (CFMEU, Mining & Energy, state secretary), Tom Price, 23 June 2004.

33 *Workforce*, 30 August, 6 September, 15 November 2002.

34 Author's interview, Tracey, Newman, 4 December 2003.

35 Pilbara Sustainability Taskforce, Minutes of Community Meeting, 26 February 2003; Media Release, 1 August 2003; Media Release, 29 October 2003; author's interview, Sharon Thiel (PST), Newman, 3 December 2003.

36 Author's interview, Thiel.

37 Author's interview, Tracey, 4 December 2003; observation, BHP-BIO Convenors Meeting, 3 December 2003.

38 Author's observation, Convenors Meeting, Newman, 3 December 2003.

39 BHP-PMU, 'Agreement in relation to unionisation of workers at BHP Iron Ore', n.d., copy kindly made available by Will Tracey, December 2003.

40 *CFMEU and others v BHP Billiton Iron Ore*, WAIRC, 12063, 2004, 12063, 20 July 2004.

41 ibid, paragraph 115.

42 ibid, paragraph 122.

43 ibid, paragraph 97.

44 ibid, paragraph 63.

45 ibid, paragraph 77.

46 In a final twist to the story of the PMU, Tracey stood for election as state secretary of the AWU in 2005 – and almost won. He subsequently went to work for the Maritime Union of Australia and in 2015 was elected Deputy National Secretary; author's interview, Tracey, Perth, 23 November 2006.

47 Author's observations of meetings; Vic Davis, personal correspondence with author, 14 August 2004.

Notes to Chapter Nine – Workers in the Boom

1 Port Jackson Partners, *Iron Ore: The Bigger Picture*, A Policy Paper Commissioned by the Minerals Council of Australia, Minerals Council of Australia, Forrest, July 2015, p. 9.

2 *Western Australian Mining and Petroleum, Statistics Digest, 2014–15,* Department of Mining and Petroleum, Western Australian Government, 2015.

3 Port Jackson Partners, *Iron Ore: The Bigger Picture*, p. 18.

4 Reserve Bank of Australia, *Statement on Monetary Policy – May,* Box C, RBA, Canberra, 2012; Australian Bureau of Statistics, cat. no. 8155.0, *Australian Industry, 2011–12*, 28 May, ABS, Canberra 2013.

5 The most comprehensive critique was in D. Richardson and R. Denniss, *Mining the Truth: The Rhetoric and Reality of the Commodities Boom*, paper no. 7, Australia Institute, 2011.

6   See several chapters in M. Brueckner et al. (eds), *Resource Curse or Cure.*

7   Unless otherwise shown, the material here is sourced from A. Burrell, *Twiggy: The High-Stakes Life of Andrew Forrest,* Black Inc., Collingwood, 2013.

8   Most recently alluded to in T. Winton, *Island Home,* Hamish Hamilton, 2015, p. 164; television footage can be readily sourced online.

9   Burrell, *Twiggy,* pp. 224, 231.

10 FMG data taken from its website: http://fmgl.com.au (accessed 30 May 2016).

11 Burrell, *Twiggy,* p. 221.

12 Summed up in P. Cleary, *Too Much Luck: The Mining Boom and Australia's Future,* Black Inc., Collingwood, 2011, pp. 75–83.

13 M. Benns, *Dirty Money: The True Cost of Australia's Mineral Boom,* William Heinemann, North Sydney, 2011.

14 ibid, chapter 7, for a comprehensive overview of all these elements.

15 On violence, see K. Carrington, A. McIntosh and J. Scott, 'Globalization, frontier masculinities and violence: Booze, blokes and brawls', *British Journal of Criminology,* (50)9, 2010, pp. 393–413.

16 Benns, *Dirty Money,* chapter 7. On drugs, working conditions and FIFO life generally, see two very funny and insightful books by Xavier Toby: *Mining My Own Business,* UWA Publishing, Crawley, 2013; *Going out of My Mined,* UWA Publishing, Crawley, 2015.

17 Author's observations of meetings and general discussions in Mt Newman, June 2010; see also Knox, *Boom,* pp. 348–9.

18 Among many overviews, see R. Cooper and B. Ellem, 'The neo-liberal state, trade unions and collective bargaining in Australia', *British Journal of Industrial Relations,* 46(3), 2008, pp. 532–54.

19 B. Pocock, J. Elton, A. Preston, S. Charlesworth, F. MacDonald, M. Baird, R. Cooper and B. Ellem, 'The impact of *Work Choices* on women in low paid employment in Australia: A qualitative analysis', *Journal of Industrial Relations,* 50(3), 2008, pp. 475–88.

20 Although many issues were important to voters, Work Choices was the vote-*changing* issue: K. Muir, *Worth Fighting For: Inside the 'Your Rights at Work' Campaign,* UNSW Press, Sydney, 2008; B. Spies-Butcher and S. Wilson, 'When labour makes a difference: Union mobilization and the 2007 federal election in Australia', *British Journal of Industrial Relations* 49(2), 2011, pp. 306–31.

21 A. Forsyth, 'The impact of "Good Faith" obligations on collective bargaining practices and outcomes in Australia, Canada and the USA', *Canadian Labour and Employment Law Journal* 16(1), 2011, pp. 1–50.

22 *Pilbara United News,* 3: 1 August 2010; for BHP Billiton, see *Common Cause,* 'Banned from using new AWAs – BHP trying a con job in the Pilbara', 74(4), 2008, p. 7.

23 Author's interviews, Dennis Jones, Karratha, 28 March 2012; car examiner, Karratha, 17 September 2011.

24 AIRC, *Coal & Allied Operations Pty Ltd v CFMEU*, Print P6645, 1372, 7 November 1997, Australian Industrial Relations Commission, 1999; *Common Cause*, 'Rio's coal war legacy is a strong union at Hunter Valley No. 1', 73(6), 2007, pp. 6–7; Mackinnon, *Behind WorkChoices*.

25 'Union revival kicks off at Rio's Pilbara miners', *Common Cause*, 74(1), 2008, p. 7; *Karratha Lodge Newsletter*, CFMEU, 1, 15 March, 2008.

26 *Pilbara United News*, CFMEU, 1/12, 30 January 2012.

27 'Union revival kicks off', *Common Cause*.

28 *Karratha Lodge Newsletter*, 3, 28 July 2008.

29 ibid.

30 Author's interviews, Jones; anonymous drivers A, B, Karratha, 27 March 2012.

31 *Karratha Lodge Newsletter*, 3, 28 July 2008.

32 Author's interviews with all leading officials repeatedly attested to this; for the public record, see *Karratha Lodge Newsletter*, various issues.

33 *Karratha Lodge Newsletter*, 3, 28 July 2008.

34 Author's interviews, Jones, Gary Wood, Karratha, 16 September 2011.

35 *Pilbara United News*, 2/09, 20 April 2009.

36 *Karratha Lodge Newsletter*, 3, 28 July 2008.

37 *Karratha Lodge Newsletter*, 4, 22 August 2008; 'Big win for Rio's Pilbara union mineworkers', *Common Cause*, 74(5), 2008, p. 11.

38 *Karratha Lodge Newsletter*, 4, 22 August 2008; *CFMEU v Pilbara Iron Company (Services) Pty Ltd, Hamersley Iron Pty Ltd and Robe River Iron Mining Co Pty Ltd*, 8210, Fair Work Australia 2010; author's interview, Wood.

39 'Big win for Rio's Pilbara Union mineworkers', *Common Cause*.

40 A. Buckley-Carr, 'Unions in Pilbara push after new mine death', *The Australian*, 28 August 2008, p. 8; P. Taylor, 'Pilbara workers to strike', *The Australian*, 7 October 2008, p. 1; P. Taylor, 'Mining union ready to dig in', *The Australian*, 8 October 2008, p. 8.

41 Quoted in P. Taylor, 'Strikers claim $14m blow to Rio Tinto', *The Australian*, 3 November 2008, p. 6.

42 ibid; 'Rio refuses to meet Pilbara Union loco drivers on new EA', *Common Cause*, 74(6), 2008, p. 6.

43 Forsyth, 'The impact of "Good Faith" obligations'.

44 *Pilbara United News*, 2/09, 20 April 2009.

45 ibid; *Pilbara United News*, 4/09, 24 September 2009.

46 *Pilbara United News*, 2/09, 20 April 2009; *Pilbara United News*, 3/09, 26 June 2009.

47 *Construction, Forestry, Mining and Energy Union v Pilbara Iron Company (Services) Pty Ltd*, Federal Court of Australia, 822, 5 August 2010.

48 'Union revival at Rio's mines', *Common Cause*, 77(4), 2010, p. 7; author's interviews, Jones, Wood; 'Pilbara mineworkers demanding the right to collective bargaining', *Common Cause*, 75(6), 2009, p. 6; *Pilbara United News*, 1/10, 9 February 2010.

49   AAP, 'Rio union deal shows a new IR era', Anonymous, AAP Bulletin Wire, Sydney, 12 January, 2010; M. Stevens, 'Dangers in union power revival', *Weekend Australian*, 6 February 2010, p. 25.

50   *Pilbara United News*, 1/10, 9 February 2010; 'Pilbara union members on verge of historic breakthrough with Rio', *Common Cause*, 76(3), 2010, p. 11.

51   'Pilbara Union members on verge of historic breakthrough with Rio', *Common Cause*.

52   'Union revival at Rio's mines', *Common Cause*, 77(4), 2011, p. 7; FWA, *Pilbara Iron Locomotive Drivers and Car Examiners Agreement*, Fair Work Australia, 6242, 2011.

53   FWA, *CFMEU v Pilbara Iron Company (Services) Pty Ltd, Hamersley Iron Pty Ltd and Robe River Iron Mining Co Pty Ltd*. Fair Work Australia 8210, paragraph 14.

54   'Gillard govt must act as Pilbara victims of Fair Work quirk', *Common Cause*, 76(6), 2010, p. 7.

55   FCAFC, *Construction, Forestry, Mining and Energy Union v Pilbara Iron Company (Services) Pty Ltd* Federal Court of Australia, Full Court, 91, 25 July 2011.

56   CFMEU, 'High Court delivers a win for collective bargaining at Rio Tinto', available at: http://www.cfmeu.com.au/high-court-delivers-a-win-for-collective-bargaining-in-rio-tinto's-pilbara-operation (accessed 17 February 2012).

57   Rio Tinto, 'Rio Tinto invests US$518 million in autonomous trains for Pilbara iron ore rail network in Western Australia', available at: http://www.riotintoironore.com/ENG/media/38_media_releases_2358.asp (accessed 5 September 2013); author's interview, Jones.

58   B. Pini and R. Mayes, 'Gender, emotions and fly-in fly-out work', *Australian Journal of Social Issues*, 47(1), 2012, pp. 71–86; R. Mayes, 'Fly-in fly-out work: Temporal and spatial re-integrations of work, family and community', 5th International Community, Work and Family Conference, University of Sydney, July 2013.

59   'CFMEU ends union drought at Rio Tinto's Pilbara operations', *Workplace Express*, 9 June 2011.

Notes to Chapter Ten – Beyond the Boom

1   J. Thomson, 'Atlas Iron to shut mining operations due to price plunge', *Sydney Morning Herald*, 11 April 2015, p. 4.

2   For background on both, see Benns, *Dirty Money*, chapter 2.

3   ibid, p. 38.

4   Knox, *Boom*, p. 318.

5   As reported by ABC news, 5 September 2012; see http://www.abc.net.au/news/2012-09-05/rinehart-says-aussie-workers-overpaid-unproductive/4243866 (accessed 26 April 2016).

6   *Sydney Morning Herald*.

gina-rineharts-roy-hill-mine-exports-first-ore-to-asia-20151208-gliqdg.html (accessed 26 April 2016).

7   BMI Research, *Australia Mining Report Q3*, Business Monitor International, London, 2015, p. 69.

8   BHP Billiton Limited, *Resourcing Global Growth Annual Report*, 2015; Rio Tinto Limited, *2014 Annual Report*, 2015; Rio Tinto Limited, *First Quarter Operations Review*, 2016.

9   P. Ker, 'The iron ore mega storm that wasn't', *Sydney Morning Herald*, 21 April 2016, p. 26

10 P. Todd, B. Ellem, C. Goods, A. Rainnie and L. Smith, 'Labour in global production networks: Workers and unions in mining engineering work', *Economic and Industrial Democracy*, 2017, DOI: 10.1177/0143831X16684964.

11 S. Morris, 'Legislation mooted over Rio's squeeze on contractors', *West Australian*, 14 April 2016.

12 BMIResearch, *Australia Mining Report Q3*, p. 25.

13 B. Witteveen, *Resources and Energy Quarterly*, edited by Department of Industry and Science, Commonwealth of Australia, Canberra, 2015; Ker, 'The iron ore mega storm that wasn't'.

14 International Longwall News, available at: http://www.longwalls.com/storyView.asp?StoryID=9589335 (accessed 19 July 2012).

15 Rio Tinto, 'Rio Tinto invests'.

16 Author's interview, Jones, Karratha, 19 May 2012.

17 ABC, 'Technology set to change face of mining boom', *7.30 Report*, Australian Broadcasting Corporation, 21 February 2012.

18 Rio Tinto, 'Rio Tinto invests'.

19 ABC, 'Technology set to change face of mining boom'.

20 Peter Strachan, ABC, 'Technology set to change face of mining boom'.

21 Sam Walsh quoted in P. Roberts, 'Drilling down', *Boss/Australian Financial Review*, 13 October 2012, p. 19.

22 Quoted in D. Kitney, 'Union push could wreck productivity, miner warns', *Australian*, 26 October 2011, p. 6.

23 Author's interviews, Jones, 19 May 2012; anonymous drivers, September 2011, as set out in chapter 9. This section relies on a shift spent on the loco in May 2012 with Denis 'Sharkey' Jones, the President of the CFMEU's Karratha Lodge. I am grateful to him for organising this and to Rio's rail supervisors for allowing it.

24 Observations in Karratha rail depot crib room, 19 May 2012; see also Knox, *Boom*, p. 305.

25 Author's interview, Jones, 19 May 2012.

26 ibid.

27 ibid.

28 'Pilbara Union members on verge of historic breakthrough with Rio', *Common Cause*, 76(3), 2010, p. 11; author's interview, Jones, 28 March 2012.

29 Rio Tinto, *Autohaul*, n.d. (fact sheet for employees; in possession of author).

30 'Robot train stuck at the station', *Sydney Morning Herald*, 20 June 2016; author's interview, Wood, Paraburdoo, 8 May 2016.

31 C. Brigden and S. Kaine, 'Rethinking factional alliances and union renewal: Inter-union collaboration in the 21st century', *Economic and Industrial Democracy*, 36(2), pp. 239–57.

32 'Unions join forces in bid to eat Rio "elephant"', *Workplace Express*, 22 May 2013.

33 Author's interview, Wood, 8 May 2016; author's observations, WMWA meetings, Paraburdoo, 8 May 2016, Tom Price, 9 May 2016.

34 WMWA pamphlet/membership form, 2016; author's observations, WMWA meetings, Paraburdoo, 8 May 2016, Tom Price, 9 May 2016.

35 Author's interviews, Wood, 8 May 2016, Paraburdoo; Shane Roulstone and Andrew Smith, WMWA organisers, Tom Price, 9 May 2016; author's observations, WMWA meetings, Paraburdoo, 8 May 2016, Tom Price, 9 May 2016.

36 Author's observations, WMWA meetings, Paraburdoo, 8 May 2016, Tom Price, 9 May 2016.

37 F. Haslam McKenzie, 'Fly-in fly-out: The challenges of transient populations in rural landscapes' in G. Luck, R. Black and D. Race (eds), *Demographic Change in Australia's Rural Landscapes*, Springer, Heidelberg, 2010, p. 357.

38 Pini and Mayes, 'Gender, emotions and fly-in fly-out work'.

How this Book was Written

The research for this book was based on seventeen research trips to the Pilbara and others to Perth, with scores of interviews, a series of focus groups, informal conversations, attendance at union meetings, observation of the November 2002 'organising blitz' in Paraburdoo, a twelve-hour shift out of Rio Tinto's Eight-Mile Depot near Karratha and several hours spent in the galleries of various tribunals and courtrooms. The book also draws on union records and publicly available material from the media, companies, the Western Australian Industrial Relations Commission, Fair Work Australia and its prior bodies and the Federal Court. Because my own direct experience of the Pilbara began in 2001, all the chapters from then on (from chapter seven) draw most heavily on my own original research and contacts. For the earlier period, I think I uncovered just about all the written work. I acknowledge the research done by many people before me. In the early chapters I have also drawn on extensive interviews undertaken in 1990-91 by Ray Fells and Stuart Reid. I am very grateful to Ray, Rob Lambert and professional staff in the Business School at the University of Western Australia for getting the original transcripts to me.

The research methods I used are explained fully in my chapter called 'Drinking with Dessie: Research, mines and life in the Pilbara', in Keith Townsend and John Burgess (eds), *Method in the Madness: Research Stories You Won't Find in a Textbook* (Chandos Publishing, Oxford, 2009). The fullest explanation of the theory behind this book is in 'Scaling labour: Australian unions and global mining', *Work, Employment and Society*, 20(2), 2006, pp. 369–87. An overview of the book's argument is in 'Resource peripheries and neo-liberalism: The Pilbara and the remaking of industrial relations in Australia', *Australian Geographer*, 46(3), 2015, pp. 323–37.

One other important thing needs to be said about the writing of this book. Almost all my own interviews were with workers or unionists. This is because when I first went to the Pilbara it was eighteen months into BHP's move to reduce union presence at its sites. It was a tense and often bitter time not only between unions and managers but between union members and those who had signed individual contracts. I thought I could write about the company's angle on this episode based on publicly available sources. It seemed to me that to win trust on the ground I had to work on one side of the fence. I would not recommend this to all researchers but it felt like all I could do at the time. Several former and current senior managers from BHP, Hamersley Iron and Robe River did speak to me, mostly off record, but that was later. I also gained much from an invitation from the Australian Mines and Metals Association in Adelaide in 2009 to speak at their annual conference. I thank Minna Knight (now a member of the Queensland Industrial Relations Commission) for this.

Like any piece of research, this book is a collective effort. I gained a lot from feedback from academic colleagues during many seminars and conferences where I presented papers on different aspects of the Pilbara saga. At the end, Trish Todd, Chris F. Wright and Al Rainnie (who told me years ago 'to just write the bloody thing') read the whole manuscript and Ray Fells lent his unique expertise to one particularly troublesome chapter. Trish also organised an honorary position for me at the University of Western Australia which was very helpful. Marian Baird provided the original suggestion about getting into the Pilbara ASAP and, with our department's great mentor, Russell Lansbury, read parts of the manuscript. I thank them all, along with Rae Cooper who provided the initial union introductions I needed back in 2001 and Janis Bailey who seemed to know everyone I needed to meet in Perth.

The full list of people who have helped me over the fifteen years of research is a long one: Tom Barratt, Tony Beech, Alex Bukarica, Sally Cawley, Andrea Chadwick, Robin Chapple, Sharon Clarke, Amanda Coles, Tony Cooke, Dave Cooper, Michael Crosby, Justine Evesson, Peter Fairbrother, Scott Fitzgerald, Neil Flynn, Jon Ford, Barney Fyfe, Mike Gillan, Caleb Goods, Paddy Gorman, Aaron Greenhalgh, Jack

Gregor, Tony Hall, Bob Horstman, Warren Johncock, Kevin Jones, Rob Lambert, Mike Llewellyn, Martie Lowenstein, Bruce Hearn Mackinnon, Tony Maher, Robyn Mayes, Stephanie Mayman, Kelvin and Ann-Marie McCann, Shannon O'Keeffe, Bobbie Oliver, Bob Powell, Mike Rafferty, Derek Schapper, Tony Slevin, Herb Thompson, Julie Tracey, Alexis Vassiley and Alex Veen. 'In the east', Trish Donaldson patiently endured my disappearances to the Pilbara and then into the study and my obsession with all things West Australian.

I received a lot of assistance in the search for photographs from the staff of the State Library of Western Australia, especially the patient and highly efficient Helen Oufs. In the Western Australian Department of State Development, Joanne Marchioro was an invaluable help in organising permissions to use the images, as were Karen Williams and Cyrus Irani at the Australian Broadcasting Corporation. I acknowledge the support of the University of Sydney's Sesquicentenary Grant Scheme and School of Business research funds in making some of the early travel possible.

My greatest debt is to the many people in (or once in) the Pilbara who provided ideas and arguments, told their own stories, sent me material and allowed me into their homes and meetings. The ACTU's Pilbara organisers, Troy Burton, Stewart Edward and Will Tracey, were extremely generous as was Gary Wood, who was the state secretary of the CFMEU's Mining & Energy Division throughout my research, and Jimmy Murie, who had worked in the Pilbara prior to becoming an official with the ETU. As with my earlier published work, thanks to Nancy Missler for the poem (I got the acknowledgment right this time, Nancy), Des for the party in Newman and Batman for that truck ride. Special thanks to Sharkey for the whole train experience in 2012. Several interviewees remain anonymous – many of the PMU members in Paraburdoo, rail workers in Karratha and union delegates in Port Hedland and Newman from 2001 on – but you know who you are and I thank you sincerely.

To the key people whom I formally interviewed, I record my deep gratitude: Lea Anderson, 'Curly' Asplin, Maria Boyington, Danny Connors, Brett Davis, Kathleen Galvin, Colin Gilbert, Graeme Haynes,

Ross Kumerca, Kerrianne and Andy Mills, Dave McLane, Colleen and Terry Palmer, Pikey, Cindy and Kelvin Portland, Kevin Quill, John Radford, Phil Randal, Shane Roulstone, Andrew Smith, Sharon Thiel, Steve and Lorna Thomas, Mark Troy. I want to make special mention of Bruce Bonnor, Vic Davis and John Johnston, who were all 'true believers' in their unions and who died too young. 'JJ' and his partner Sharon were vital in linking me up with everyone I needed to meet in Newman; good cook, too, was JJ.

Turning all this work into a book needed a publisher willing to give it a go and that person was the tireless and wonderful Terri-ann White at UWA Publishing. Once on board, I was helped immensely by the very capable and good-humoured Kate Pickard and Charlotte Guest; it was an absolute delight to work with them along with Tracy Peacock in publicity. Suzannah Shwer did a fabulous job of editing the manuscript, turning some real clunkers into much better prose. Thanks to you all. Obviously, with all this help, any flaws in the book must be sheeted home to me.

Permissions

I am grateful to the Australian Society for the Study of Labour History and *Labour History*'s editor, Professor Diane Kirkby, for permission to reproduce revised sections of my article on Robe River ('Robe River Revisited: Geohistory and Industrial Relations', *Labour History*, 109, 2015, pp. 111–30) in chapter 5. Sage allows its journal authors to reproduce sole-authored work without permissions. Revised versions of parts of these papers in Sage journals were used: 'Scaling Labour: Australian Unions and Global Mining', *Work, Employment and Society*, 20 (2), 2006, pp. 369–87 (chapters 7 and 8); 'A Battle Between Titans? Rio Tinto and Union Recognition in Australia's Iron Ore Industry', *Economic and Industrial Democracy*, 35 (1), 2014, pp. 185–200 (chapter 9); 'Geographies of the Labour Process: Automation and the Spatiality of Mining', *Work, Employment and Society*, 30 (6), 2016, pp. 932–48 (chapter 10).

All the images in the book, including the cover image, were sourced from the State Library of Western Australia. Two sets of images required permissions for reproduction: for images 15 and 16, I am grateful to the ABC; for images 2, 3 and 4, I acknowledge the Department of State Development.

References

Author's interviews and observations

Focus groups
Paraburdoo, Tom Price, Dampier, Karratha, 24–29 June, 2002.

Anonymous interviews
BHPIO delegates, Newman and Port Hedland, 26–29 June,
    17, 20 September 2001.
Car examiner, Karratha, 17 September 2011.
Hamersley employees, various, Tom Price, 18 September 2001.
'John', train driver, Port Hedland, 29 June 2001.
Loco driver A, Karratha, 27 March 2012.
Loco driver B, Karratha, 27 March 2012.
Small business owners, Paraburdoo, 16 November 2002.

Named interviews
Lea Anderson (former Pilbara AMWU administrative worker), Perth,
    27 November 2006.
Bruce Bonnor (former Robe River employee), Mt Newman, 16 November
    2005.
Maria Boyington (Action in Support of Partners), South Hedland, 29 June
    2001.
Troy Burton (ACTU organiser), Sydney, 16 May 2002; Paraburdoo,
    15 November 2002.
Stewart Edward (ACTU organiser), Paraburdoo, 16 November 2002.
Kathleen Galvin (ACTU organiser), Paraburdoo, 15 November 2002.
Graeme Haynes (former ETU convenor), Perth, 23 November 2006.
Dennis Jones (Karratha Lodge President CFMEU), Karratha, 28 March
    2012, 19 May 2012.
Ross Kumeroa (convenor), Newman, 26 June 2001.
Dave McLane (former CMEU organiser), 23 November 2006, Perth;
    phone, 20 June 2016.
Jim Murie (ETU official), Paraburdoo, 16 November 2005.

Colleen Palmer (Action in Support of Partners), South Hedland, 28 June 2001.
Phil Randal (former Hamersley employee), Blacksmiths, 11 August 2004.
Shane Roulstone (WMWA organiser), Tom Price, 9 May 2016.
Derek Schapper (union counsel), Perth, 18 October 2001.
Andrew Smith (WMWA organiser), Tom Price, 9 May 2016.
Will Tracey (ACTU organiser), Newman, 27 June 2001; Newman, 17 September 2001; Karratha, 19 September 2001; en route Tom Price–Dampier, 27 June 2002; Dampier, 27 June 2002; Karratha, 28 June 2002; Newman, 4 December 2003; Paraburdoo 23 June 2004; Karratha, 26 June 2004; Perth, 23 November 2006.
Sharon Thiel (Pilbara Sustainability Taskforce), Newman, 3 December 2003.
Mark Troy (former Hamersley employee), Karratha, 6 December 2003.
Gary Wood (CFMEU, Mining and Energy Division, state secretary), Tom Price, 23 June 2004; Perth, 16 September 2011; 8 May 2016, Paraburdoo.

Observations
BHP convenors meeting, Newman, 3 December 2003.
Delegates meetings, Newman and Port Hedland, 26–29 June 2001.
Eight Mile Depot, Karratha, crib room, 19 May 2012.
Locomotive, Eight Mile Depot–Tom Price rail-line, 19 May 2012.
Paraburdoo blitz meetings, 15 November 2002.
PMU committee, Paraburdoo, 4 December 2003.
PMU committees, Tom Price, Paraburdoo, Karratha, 23–25 June 2004.
WMWA meetings, Paraburdoo, 8 May 2016, Tom Price, 9 May 2016.

Bibliography

AAP, 'Rio union deal shows a new IR era', Anonymous, AAP Bulletin Wire, Sydney, 12 January, 2010.
ABC News, 'Aussies must compete with $2 a day workers: Rinehart', 5 September 2012; available at: http://www.abc.net.au/news/2012-09-05/rinehart-says-aussie-workers-overpaid-unproductive/4243866.
ABC, 'Technology set to change face of mining boom', *7.30 Report*, Australian Broadcasting Corporation, 21 February 2012.
ABS, *Employee Earnings, Benefits and Trade Union Membership, Australia*, cat. no. 6310.0, Australian Bureau of Statistics, Canberra, various years.

ABS, *Australian Industry, 2011–12*, cat. no. 8155.0, Australian Bureau of Statistics, Canberra 2013.

ACTU, 'BHPIO – Port Hedland Campaign Planning Sessions, internal notes, 10–12 December 1999.

ACTU, *unions@work*, ACTU, Melbourne, 1999.

AIRC, *Coal & Allied Operations Pty Ltd v CFMEU*, Print P6645, [1999] 1372, 7 November, Australian Industrial Relations Commission, 1997.

AIRC, *Australian Workers Union and Hamersley Iron; Construction Forestry, Mining and Energy Union & others, Decision*, PR947647, Australian Industrial Relations Commission, 2004.

Allen, R., 'The End of the Closed Shop', *Proceedings of the Conferences of the HR Nicholls Society*, vol. 13, 1992, available at: http://archive.hrnicholls. com.au/archives/vol13/vol13-5.php (accessed 24 November 2014).

Allen, R., 'Agreements without Unions', *Proceedings of the Conferences of the HR Nicholls Society*, vol. 15, 1994, available at: http://archive.hrnicholls. com.au/archives/vol15/vol15-7.php (accessed 2 March 2016).

Australian Mining Monthly, September, 2011.

AWU, *Information and Documents Concerning the Federal Award Developments at Hamersley Iron & Robe River*, Australian Workers' Union, Melbourne, 2003.

Bachelard, M., 'The day a myth called happiness died', *Weekend Australian*, 6–7 April 2002.

Bailey, J. and Hortsman, B., '"Life is full of choices": Industrial relations "reform" in Western Australia since 1993', Association of Industrial Relations Academics of Australia and New Zealand, 14th Conference, Newcastle, February 2000.

Bain, G. S. and Price, R., *Profiles of Union Growth: A Comparative Statistical Portrait of Eight Countries*, Basil Blackwell, Oxford, 1980.

Balogh, S., 'Back in business', *Weekend Australian*, 26–7 July 2003.

BCA, *Enterprise-Based Bargaining Units: A Better Way of Working*, Business Council of Australia Melbourne, 1989.

Benns, M., *Dirty Money: The True Cost of Australia's Mineral Boom*, William Heinemann, North Sydney, 2011.

Berndt, R. M., 'Traditional Aboriginal life in Western Australia: as it was and is', in R. M. and C. H. Berndt, *Aborigines of the West: Their Past and Their Present*, University of Western Australia Press, Nedlands, 2nd edition, 1980, pp 3–27.

BHP Billiton Limited, *BHP Billiton Resourcing Global Growth Annual Report*, 2015. Available at: http://www.bhpbilliton. com/~/media/bhp/documents/investors/annual-reports/2015/ bhpbillitonannualreport2015_interactive.pdf?la=en.

BHP Iron Ore, *Staff Handbook*, copies made available to author, 2001.

BHPPMU, Agreement in relation to unionisation of workers at BHP Iron Ore, n.d., copy kindly made available to author by Will Tracey, December 2003.

Bickerton, M., *Dust Over the Pilbara*, Artlook, Perth, 1980.

Blackburn, E., *Broken Lives*, Hardie Grant Books, South Yarra, 2001.

Blainey, G., *The Rush That Never Ended: A History of Australian Mining*, Melbourne University Press, Carlton, 2nd edition, 1969.

BMI Research, *Australia Mining Report Q3*, Business Monitor International, London, 2015.

Bolton, G., *Land of Vision and Mirage: Western Australia Since 1826*, University of Western Australia Press, Crawley, 2008.

Braverman, H., *Labor and Monopoly Capital*, Monthly Review Press, New York, 1974.

Brigden, C., 'Power and space in the Victorian Trades Hall Council', in B. Ellem, R. Markey and J. Shields (eds), *Peak Unions in Australia: Origins, Purpose, Power, Agency*, The Federation Press, Leichhardt, 2004, pp. 219–35.

Brigden, C. and Kaine, S., 'Rethinking factional alliances and union renewal: Inter-union collaboration in the 21st century', *Economic and Industrial Democracy*, 36(2), pp. 239–57.

Brown, W., 'The influence of product markets on industrial relations', in P. Blyton, N. Bacon, J. Fiorito and E. Heery (eds), *The Sage Handbook of Industrial Relations*, Sage, London, pp. 113–28.

Brueckner, M., Durey, A., Mayes, R. and Pforr, C. (eds), *Resource Curse or Cure: On the Sustainability of Development in Western Australia*, Springer, Heidelberg, 2014.

Buckley-Carr A., 'Unions in Pilbara push after new mine death', *The Australian*, 28 August 2008.

Bulbeck, C., 'The iron ore stockpile and dispute activity in the Pilbara', *Journal of Industrial Relations*, 25(4), 1983, pp. 431–44.

Burrell, A., *Twiggy: The High-Stakes Life of Andrew Forrest*, Black Inc., Collingwood, 2013.

Carrington, K., McIntosh, A. and Scott, J., 'Globalization, frontier masculinities and violence: Booze, blokes and brawls', *British Journal of Criminology*, (50)9, 2010, pp. 393–413.

CFMEU, *High Court Delivers a Win for Collective Bargaining in Rio Tinto*, 2012, available at: http://www.cfmeu.com.au/high-court-delivers-a-win-for-collective-bargaining-in-rio-tinto%E2%80%99s-pilbara-operation (accessed 17 February, 2012).

Cleary, P., *Too Much Luck: The Mining Boom and Australia's Future*, Black Inc., Collingwood, 2011.

Common Cause various issues, CFMEU, Mining and Energy Division, Sydney, copies in possession of author.

Cooper, R., 'Organise, organise, organise! ACTU Congress 2000', *Journal of Industrial Relations* 42(4), 2000, pp. 582–94.

Cooper, R. and Ellem, B., 'Union power: Space, structure, strategy', in M. Hearn and G. Michelson (eds), *Rethinking Work: Time, Space and Discourse*, Cambridge University Press, Cambridge, 2006, pp. 123–43.

Cooper, R. and Ellem, B., 'The neo-liberal state, trade unions and collective bargaining in Australia', *British Journal of Industrial Relations*, 46(3), 2008, pp. 532-54.

Copeman, C., 'The Robe River Affair', H. R. Nicholls Society Archives, 1987, available at: http://archive.hrnicholls.com.au/archives/vol3/vol3-8.php (accessed 5 November 2013).

Copeman, C. 'I.R. Lessons from Recent Mining Industry History', H. R. Nicholls Society Archives, 1992, available at: http://archive.hrnicholls.com.au/archives/vol13/vol13-7.php, (accessed 5 November 2013).

Court, H., 'Industrial Disputes in the Pilbara', unpublished MEc thesis, University of Western Australia, 1976.

Coyne, M. and Edwards, L., *The Oz Factor: Who's Doing What in Australia*, Dove Communications, East Malvern, 1980.

Crosby, M., *Power at Work: Rebuilding the Australian Union Movement*, The Federation Press, Leichhardt, 2005.

Dabscheck, B., *Australian Industrial Relations in the 1980s,* Oxford University Press, Melbourne, 1989.

Dabscheck, B. and Niland, J., *Industrial Relations in Australia*, George Allen & Unwin, North Sydney, 1985 (first published 1981).

Davis, V., personal communication with author, 14 August 2004.

Day, D., Collection of personal material deposited in union offices in Port Hedland, various dates, copies kindly made available to the author.

Deery S. and Mitchell, R. (eds), *Employment Relations: Individualisation and Union Exclusion*, The Federation Press, Annandale, 1999.

Dufty, N., 'The 1979 Hamersley strike', *Australian Bulletin of Labour*, 8(4), 1982, pp. 210–28.

Dufty, N., *Industrial Relations in the Pilbara Iron Ore Industry,* Western Australian Institute of Technology, Perth, 1984.

Dufty, N. 'Remoteness and industrial relations – the Pilbara iron ore industry', in T. B. Brealey, C. C. Neil and P. W. Newton (eds), *Resource Communities: Settlement and Workforce Issues*, CSIRO Publishing, Melbourne, 1988, pp. 275–89.

Edward, S., personal communication with author, 17 April 2003.

Edwards, R., *Contested Terrain: The Transformation of the Workplace in the Twentieth Century*, Basic Books, New York, 1979.

Eklund, E., *Mining Towns: Making a Living, Making a Life*, New South Publishing, Sydney, 2012.

Ellem, B., 'Trade unionism in 1998', *Journal of Industrial Relations*, 41(1), 1999, pp. 127–51.

Ellem, B. and Shields, J., 'Making a "union town": Class, gender and workers' control in inter-war Broken Hill', *Labour History*, 78, 2000, pp. 116–40.

FCA, *Australian Workers' Union v BHP Iron Ore Limited*, Federal Court of Australia, 430, 2000.

FCA, *Australian Workers' Union v BHP Iron Ore Limited*, Federal Court of Australia, 3, 2001.

FCA, *Construction, Forestry, Mining and Energy Union v Pilbara Iron Company (Services) Pty Ltd*, Federal Court of Australia, 822, 5 August 2010.

FCAFC, *Construction, Forestry, Mining and Energy Union v Pilbara Iron Company (Services) Pty Ltd*, Federal Court of Australia, Full Court, 91, 25 July 2011.

Fells, R., 'Award restructuring – but where did we get our awards in the first place? The development of the award and union structures in the iron ore industry', in C. Fox (ed.), *Papers in Labour History 11*, Australian Society for the Study of Labour History, Perth, 1993, n.p.

Fells, R. and Reid, S., '"Tons were the go": Some recollections of work in the early days of the Pilbara iron ore industry', in C. Fox and M. Hess (eds), *Papers in Labour History 8*, Australian Society for the Study of Labour History, Perth, 1991, pp. 1–16.

Flanders, A., *Management and Unions: The Theory and Reform of Industrial Relations*, Faber and Faber, London, 1970.

Ford, J., 'Unions in the Pilbara', Parliament of Western Australia, Legislative Council, 22 August 2001, copy kindly made available by Jon Ford.

Forsyth, A., 'The impact of "Good Faith" obligations on collective bargaining practices and outcomes in Australia, Canada and the USA', *Canadian Labour and Employment Law Journal*, 16(1), 2011, pp. 1–50.

Frenkel, S., 'Industrial conflict, workplace characteristics and accommodation structures in the Pilbara iron ore industry', *Journal of Industrial Relations*, 20(4), 1978, pp. 386–406.

Getlin, P., *The Power Switch at Robe*, Australian Institute for Public Policy, Perth, 1990.

Gollan, R. A., *The Coalminers of New South Wales: A History of the Union, 1860-1960*, Melbourne University Press, Melbourne, 1963.

Goodrich, C. L., *The Frontier of Control: A Study in British Workplace Politics*, foreword by R. H. Tawney, 1920; republished with a new foreword and additional notes by R. Hyman, Pluto Press, London, 1975.

Graham, L., MLA, Pilbara, Western Australian Legislative Assembly, *Hansard*, 4939–95, 4 September, 1991.

Griffiths, M., *Of Mines and Men: Australia's 20th Century Mining Miracle, 1945–1985*, Kangaroo Press, East Roseville, 1998.

Hale, M., (Minyjun), *Kurlumarniny: We Come From the Desert*, Aboriginal Studies Press, Canberra, 2012.

Hamersley Iron 'Conditions of Employment', n.d; documents kindly made available to author by anonymous employees.

Hardie, J., *Nor'Westers of the Pilbara Breed*, Shire of Port Hedland, Port Hedland, 1981.

Harman, E. J. and Head, B. W. (eds), *State, Capital and Resources in the North and West of Australia*, University of Western Australia Press, Nedlands, 1982.

Haslam McKenzie, F., 'Fly-in fly-out: The challenges of transient populations in rural landscapes' in G. Luck, R. Black and D. Race (eds), *Demographic Change in Australia's Rural Landscapes*, Springer, Heidelberg, 2010, pp. 353–74.

Heath, C. and Bulbeck, C., *Shadow of the Hill*, Fremantle Arts Centre Press, Fremantle, 1985.

Heiler, K., Pickersgill, R. and Briggs, C., *Working Time Arrangements in the Australian Mining Industry: Trends and Implications with Particular Reference to Occupational Health and Safety*, ILO, Geneva, 2000.

Hess, M., 'Black and red: The Pilbara pastoral workers' strike, 1946', *Aboriginal History*, 18(1-2), 1994, pp. 65–86.

HI Process Design Project, *Implementation Manual*, Hamersley Iron, 1998[?]; documents kindly made available to author by anonymous employees.

Horsley, J., 'Conceptualising the state, governance and development in a semi-peripheral resource economy: The evolution of State Agreements in Western Australia', *Australian Geographer*, 44(3), 2013, pp. 283–303.

International Longwall News, available at: http://www.longwalls.com/storyView.asp?StoryID=9589335 (accessed 19 July 2012).

Iron Ore History Project, interviews conducted 1990–91 by Stuart Reid and Ray Fells; kindly made available by Rob Lambert.

Jobline, Robe River dispute materials kindly provided by Bruce Bonnor.

Karratha Lodge Newsletter, CFMEU, Karratha; various dates; copies in possession of author.

Kelly, P., *The End of Certainty: The Story of the 1980s*, Allen & Unwin, St Leonards, 1992.

Ker, P., 'The iron ore mega storm that wasn't', *Sydney Morning Herald*, 21 April 2016.

Kitney, D., 'Union push could wreck productivity, miner warns', *The Australian*, 26 October 2011.

Knox, M., *Boom: The Underground History of Australia, From Gold Rush to GFC*, Viking, Melbourne, 2013.

Lewis, M., *Conversations with the Mob*, University of Western Australia Press, Crawley, 2008.

Lockwood, R., *War on the Waterfront: Menzies, Japan and Pig-iron Dispute*, Hale & Iremonger, Sydney, 1987.

Lovett, E. P., 'The Structure of Industrial Negotiations within the Pilbara Iron Ore Industry', unpublished MA thesis, Monash University, 1980.

Lowenstein, W., *Weevils at Work: What's Happening to Work in Australia*, Catalyst Press, Annandale, 1997.

Ludeke, J. T., *The Line in the Sand: The Long Road to Staff Employment in Comalco*, Wilkinson Books, Melbourne, 1996.

Mackinnon, B. H., *Behind WorkChoices: How One Company Changed Australia's Industrial Relations*, Heidelberg Press, Heidelberg, 2007.

MacWilliam, 'Marginalizing opposition to the Accord', *Arena*, 92, 1990, pp. 146–50.

Massey, D., *Spatial Divisions of Labour: Social Structures and the Geography of Production*, Macmillan, London, (first published 1984) 1995.

Mayes, R., 'Fly-in fly-out work: Temporal and spatial re-integrations of work, family and community', 5th International Community, Work and Family Conference, University of Sydney, July 2013.

McRae, I., Letters to State Secretaries, 31 July, 1986; *Memo To All Staff: Union Representatives*, 31 July 1986; *Notice to All Employees*, 31 July 1986; Robe River dispute materials kindly provided by Bruce Bonnor.

Metcalf, A., *For Freedom and Dignity: Historical Agency and Class Structures in the Coalfields of NSW*, Allen & Unwin, North Sydney, 1988.

Milliss, R., *Waterloo Creek: The Australia Day Massacre of 1838, George Gipps and the British Conquest of New South Wales*, McPhee Gribble, Ringwood, 1992.

Mitchell, T., *Carbon Democracy: Political Power in the Age of Oil*, Verso, London, 2011.

Morris, S., 'Legislation mooted over Rio's squeeze on contractors', *The West Australian*, 14 April 2016.

Muir, K., *Worth Fighting For: Inside the 'Your Rights at Work' Campaign*, UNSW Press, Sydney, 2008.

Nicholls Society, H. R.. *Arbitration in Contempt. The Proceedings of the Inaugural Seminar of the H. R. Nicholls Society Held in Melbourne 28 February – 2 March, 1986*, H. R. Nicholls Society, Melbourne, 1986.

Oliver, B., *Unity is Strength: A History of the Australian Labor Party and the Trades and Labor Council in Western Australia*, API Network, Perth, 2003.

Olive, N., *Enough is Enough: A History of the Pilbara Mob*, Fremantle Arts Centre Press, Fremantle, 2007.

Peck, J., 'Excavating the Pilbara: A Polanyian exploration', *Geographical Research*, 51(3), 2013, pp. 227–42.

Peetz, D., 'Trend analysis of union membership', *Australian Journal of Labour Economics*, 8(1), 2005, pp. 1–24.

Peetz, D. and Murray, G., '"You get really old, really quick": Involuntary long hours in the mining industry', *Journal of Industrial Relations*, 53(1), 2011, pp. 13–29.

Pekobusters, Robe River dispute materials kindly provided by Bruce Bonnor.

Phillipson, N., *Man of Iron*, Wren Books, Melbourne, 1974.

Pilbara Mineworkers' Union, Membership form, 2002; copy kindly made available by Stewart Edward.

Pilbara Sustainability Taskforce, Minutes of Community Meeting, 26 February 2003.

Pilbara Sustainability Taskforce, Media Release, 1 August 2003.

Pilbara Sustainability Taskforce, Media Release, 29 October 2003.

Pilbara United News, CFMEU, Karratha, various dates; copies in possession of author.

Pini, B. and Mayes, R., 'Gender, emotions and fly-in fly-out work', *Australian Journal of Social Issues*, 47(1), 2012, pp. 71–86.

Pocock, B., Elton, J., Preston, A., Charlesworth, S., MacDonald, F., Baird, M., Cooper, R. and Ellem, B., 'The Impact of *Work Choices* on Women in Low Paid Employment in Australia: A Qualitative Analysis', *Journal of Industrial Relations*, 50(3), 2008, pp. 475–88.

Port Jackson Partners, *Iron Ore: The Bigger Picture*, A Policy Paper Commissioned by the Minerals Council of Australia, Minerals Council of Australia, Forrest, July 2015.

Priest, M., 'Union war threat after Rio deal', *Australian Financial Review*, 23 July 2003.

RBA, *Statement on Monetary Policy – May: Box C*, Reserve Bank of Australia, 2012, available at: http://www.rba.gov.au/publications/smp/2012/may/html/box-c.html (accessed 7 November 2012).

Read, J., *Marksy: The Life of Jack Marks*, Read Media, South Fremantle, 1998.

Reid, S., 'Gil Barr's story, Part 2', in C. Fox and M. Hess (eds) *Papers in Labour History 5*, Australian Society for the Study of Labour History, Perth, 1990, pp. 18–34.

Resist to Exist, Robe River dispute materials kindly provided by Bruce Bonnor.

Reynolds, T. and Dawson, V., *Iron Will: Cleveland-Cliffs and the Mining of Iron Ore, 1847-2006,* Wayne State University Press, Detroit, 2011.

Richardson D. and Denniss, R., *Mining the Truth: The Rhetoric and Reality of the Commodities Boom*, paper no. 7, Australia Institute, 2011.

Rio Tinto, *The Way We Work*, available at: http://www.riotinto.com/documents/The_way_we_work.pdf

Rio Tinto, 'Rio Tinto invests US$518 million in autonomous trains for Pilbara iron ore rail network in Western Australia', 2012, available at: http://www.riotintoironore.com/ENG/media/38_media_releases_2358.asp (accessed 5 September 2013).

Rio Tinto, Submission to the Fair Work Review, 2012, available at: http://submissions.deewr.gov.au/sites/Submissions/FairWorkActReview/Documents/RioTinto.pdf (accessed 5 September 2013).

Rio Tinto Limited, *2014 Annual report*, 2015, available at: http://www.riotinto.com/ar2014/pdfs/rio-tinto_2014-annual-report.pdf.

Rio Tinto Limited, *First quarter operations review*, 2016, available at: http://www.riotinto.com/documents/160419_Rio_Tinto_releases_robust_first_quarter_production_results_(1).pdf.

Rock Solid, various dates; copies in possession of author.

Rowley, C. D., *The Remote Aborigines*, ANU Press, Canberra, 1971.

Rowley, C. D., *The Destruction of Aboriginal Society*, Penguin, Ringwood, 1972.

Scabline, Robe River dispute materials kindly provided by Bruce Bonnor.

Scott, A., 'Looking to Sweden in order to reconstruct Australia', *Scandinavian Journal of History,* 34(3), 2009, pp. 330–52.

Smith, H. and Thompson, H., 'Industrial relations and the law: A case study of Robe River,' *The Australian Quarterly*, 59(3-4), 1987, pp. 297–305.

Snell, W., 'The Formation of a Labour Force in a Developing Region: The Case of Hamersley Iron', unpublished MEc thesis, University of Western Australia, 1969.

Spies-Butcher, B. and Wilson, S., 'When labour makes a difference: Union mobilization and the 2007 federal election in Australia', *British Journal of Industrial Relations* 49(2), 2011, pp. 306–31.

Stevens, M., 'Dangers in union power revival', *Weekend Australian,* 6 February 2010.

Storey, K., 'Fly-in/fly-out and fly-over: Mining and regional development in Western Australia', *Australian Geographer*, 32(2), 2001, pp. 133–48.

Sydney Morning Herald, 'Gina Rinehart's Roy Hill mine exports first ore to Asia', available at http://www.smh.com.au/business/mining-and-resources/gina-rineharts-roy-hill-mine-exports-first-ore-to-asia-20151208-gliqdg.html (accessed 22 April 2016).

Taylor, P., 'Pilbara workers to strike', *The Australian*, 7 October 2008.

Taylor, P., 'Mining union ready to dig in', *The Australian*, 8 October 2008.

Taylor, P., 'Strikers claim $14m blow to Rio Tinto', *The Australian*, 3 November 2008.

Thompson, E. P., *The Making of the English Working Class*, Penguin Books, Harmondsworth, 1975 (first published 1963).

Thompson, H., 'Shorter working hours or redundancies: Class conflict in the Pilbara', *Arena*, 56, 1980, pp. 113–27.

Thompson, H., 'Normalisation: Industrial relations and community control in the Pilbara', *The Australian Quarterly*, 53(3), 1981, pp. 301–24.

Thompson, H., 'The capital-labour relation in the mining sector', *Journal of Australian Political Economy*, 15, 1983, pp. 64–85.

Thompson, H., 'The pyramid of power: Transnational corporations in the Pilbara', in E. L Wheelwright and K. Buckley (eds), *Essays in the Political Economy of Australian Capitalism: Volume Five*, Australia and New Zealand Book Company, Sydney, 1983, pp. 75–100.

Thompson, H., 'Class and gender in the Pilbara', *Arena*, 68, 1984, pp. 124–40.

Thompson, H., 'The Pilbara Iron Ore Industry: Mining Cycles and Capital-Labour Relations'. *Journal of Australian Political Economy*, 21, 1987, pp. 66–82.

Thompson, H., 'The APPM dispute: The dinosaur and turtles vs the ACTU', *Economic and Labour Relations Review*, 3(2), 1992, pp. 148–64.

Thompson, H., 'The ACTU and the struggle for irrelevance', *Green Left Weekly* 7 April 1993, available at: https://www.greenleft.org.au/node/4315 (accessed 10 November 2014).

Thompson, H. and Bartlem, D., 'Confrontation in the Pilbara: The Hamersley iron ore strike', *Arena*, 55, 1980, pp. 15–31.

Thompson, H. and Smith, H., 'Conflict at Robe River', *Arena*, 79, 1987, pp. 76–91.

Thompson, P. and Macklin, R., *The Big Fella: The Rise and Rise of BHP Billiton*, Random House, North Sydney, 2009.

Thomson J., 'Atlas Iron to shut mining operations due to price plunge', *The Sydney Morning Herald*, 11 April 2015.

Toby, X., *Mining My Own Business*, UWA Publishing, Crawley, 2013.

Toby, X., *Going Out of My Mined*, UWA Publishing, Crawley, 2015.

Todd, P., Ellem, B., Goods, C., Rainnie, A. and Smith, L. (2017), 'Labour in global production networks: Workers and unions in mining engineering work', *Economic and Industrial Democracy*, DOI: 10.1177/0143831X16684964.

Tracey, J., 'The construction phase of the Pilbara iron ore industry 1965–1972: Workers, their unions, and organising the industry', *Papers in Labour History*, 13, Australian Society for the Study of Labour History, Perth, 1994, n.p.

Treadgold, T., 'Out of town and out of touch', *Business Review Weekly*, 27 June to 3 July 2002.

Trinca, H. and Davies, A., *Waterfront: The Battle That Changed Australia*, Doubleday, Milsons Point, 2000.

Tsokhas, K., *Beyond Dependence: Companies, Labour Processes and Australian Mining*, Oxford University Press, Melbourne, 1986.

Wangka Maya Pilbara Aboriginal Language Centre, http://www.wangkamaya.org.au (accessed 13 January 2016).

Western Australian Industrial Relations Commission, *BHP Iron Ore Award*, 04082, WAIRC, 2001.

Western Australian Industrial Relations Commission, *AFMEPKIUA and others vs BHPIO*, 05810, WAIRC, 2002.

Western Australian Industrial Relations Commission, *CFMEU and others v BHP Billiton Iron Ore*, 12063, WAIRC, 2004.

Western Australian Industrial Gazette.

White, O., *Under the Iron Rainbow: Northwest Australia Today*, Heinemann, Melbourne, 1969.

Wilson, D., *Different White People: Radical Activism for Aboriginal Rights 1946-1972*, UWA Publishing, Crawley, 2015.

Wilson, J., 'The Pilbara Aboriginal Social Movement: An outline of its background and significance', in R. M. and C. H. Berndt (eds), *Aborigines of the West: Their Past and Their Present*, University of Western Australia Press, Nedlands, 2nd edition, 1980.

Winton, T., *Island Home*, Hamish Hamilton, 2015.

Witteveen, B. *Resources and Energy Quarterly*, edited by Department of Industry and Science, Commonwealth of Australia, Canberra, 2015.

Workforce, various issues.

Workplace Express, 'CFMEU ends union drought at Rio Tinto's Pilbara operations', 9 June 2011.

Workplace Express, 'Unions join forces in bid to eat Rio "elephant"', 22 May 2013.

Wright, C. F., 'The Prices and Incomes Accord: Its significance, impact and legacy', *Journal of Industrial Relations*, 56(2), 2014, pp. 264–72.

Index

www.ingramcontent.com/pod-product-compliance
Lightning Source LLC
Chambersburg PA
CBHW071848270326
41929CB00013B/2148